Microbiology Experiments:
A Health Science Perspective

Fourth Edition

John Kleyn

Mary Bicknell
University of Washington

 **Higher Education**

Boston   Burr Ridge, IL   Dubuque, IA   Madison, WI   New York   San Francisco   St. Louis
Bangkok   Bogotá   Caracas   Kuala Lumpur   Lisbon   London   Madrid   Mexico City
Milan   Montreal   New Delhi   Santiago   Seoul   Singapore   Sydney   Taipei   Toronto

MICROBIOLOGY EXPERIMENTS: A HEALTH SCIENCE PERSPECTIVE
FOURTH EDITION

Published by McGraw-Hill, a business unit of The McGraw-Hill Companies, Inc., 1221
Avenue of the Americas, New York, NY 10020. Copyright © 2004, 2001, 1999, 1995
by The McGraw-Hill Companies, Inc. All rights reserved. No part of this publication may
be reproduced or distributed in any form or by any means, or stored in a database or retrieval
system, without the prior written consent of The McGraw-Hill Companies, Inc., including,
but not limited to, in any network or other electronic storage or transmission, or broadcast
for distance learning.

Some ancillaries, including electronic and print components, may not be available to customers
outside the United States.

 This book is printed on recycled, acid-free paper containing 10% postconsumer waste.

1 2 3 4 5 6 7 8 9 0 QPD/QPD 0 9 8 7 6 5 4 3

ISBN 0-07-247624-9

Publishers: *Martin J. Lange/Colin H. Wheatley*
Senior developmental editor: *Deborah Allen*
Marketing manager: *Tami Petsche*
Senior project manager: *Rose Koos*
Senior production supervisor: *Laura Fuller*
Senior media project manager: *Stacy A. Patch*
Senior media technology producer: *Barbara R. Block*
Designer: *David W. Hash*
Cover designer: *Rokusek Design*
Senior photo research coordinator: *John C. Leland*
Photo researcher: *Connie Mueller*
Compositor: *Electronic Publishing Services Inc., NYC*
Typeface: *11.5/13 Goudy*
Printer: *Quebecor World Dubuque, IA*

All experiments in this manual have been performed safely by students in college laboratories
under the supervision of the authors. Every care has been taken in the design of experiments to
avoid potentially hazardous situations. However, unanticipated events are possible due to
failure to follow proper procedures, incorrect measurement of chemicals, inappropriate use of
laboratory equipment, and other reasons. The authors and the publisher hereby disclaim any
liability for personal injury or property damage claimed to have resulted from the use of this
laboratory manual.

On the cover: When grown in extremely hard agar, the simple branching pattern exhibited
during growth of *Peanibacillus dendritiformis* changes as the organisms congregate into tight
rotating vortices that cut through the medium like a circular saw. This cooperative behavior
enables the colony to continue to grow outward. Photo provided by E. Ben-Jacob, grown by
E. Braines, photography by S. Avikam.

www.mhhe.com

To our mothers who tirelessly encouraged us.

To our spouses who supported us with their patience, suggestions, and good humor.

To our students who caught the excitement of the microbial world.

CONTENTS

Preface

To the Student

A microbiology laboratory is valuable because it actually gives you a chance to see and study microorganisms firsthand. In addition, it provides you with the opportunity to learn the special techniques used to study and identify these organisms. The ability to make observations, record data, and analyze results is useful throughout life.

It is very important to read the scheduled exercises before coming to class, so that class time can be used efficiently. It is helpful to ask yourself the purpose of each step as you are reading and carrying out the steps of the experiment. Sometimes it will be necessary to read an exercise several times before it makes sense.

Conducting experiments in microbiology laboratories is particularly gratifying because the results can be seen in a day or two (as opposed, for instance, to plant genetics laboratories). Opening the incubator door to see how your cultures have grown and how the experiment has turned out is a pleasurable moment. We hope you will enjoy your experience with microorganisms as well as acquire skills and understanding that will be valuable in the future.

To the Instructor

The manual includes a wide range of exercises—some more difficult and time-consuming than others. Usually more than one exercise can be done in a two-hour laboratory period. In these classes, students can actually see the applications of the principles they have learned in the lectures and text. We have tried to integrate the manual with the text *Microbiology: A Human Perspective*, Fourth Edition by Eugene Nester et al.

The exercises were chosen to give students an opportunity to learn new techniques and to expose them to a variety of experiences and observations. It was not assumed that the school or department had a large budget, thus exercises have been written to use as little expensive media and equipment as possible. The manual contains more exercises than can be done in one course so that instructors will have an opportunity to select the appropriate exercises for their particular students and class. We hope that the instructors find these laboratories an enjoyable component of teaching microbiology.

Acknowledgments

We would like to acknowledge the contributions of the lecturers in the Department of Microbiology at the University of Washington who have thoughtfully honed laboratory exercises over the years until they really work. These include Dorothy Cramer, Carol Laxson, Mona Memmer, Janis Fulton, and Mark Chandler. Special thanks to Dale Parkhurst for his expert knowledge of media. We also thank the staff of the University of Washington media room for their expertise and unstinting support.

We also want to thank Eugene and Martha Nester, Nancy Pearsall, Denise Anderson and Evans Roberts for their text *Microbiology: A Human Perspective*. This text was the source of much of the basic conceptual material and figures for our laboratory manual. And with great appreciation, many thanks to our editor, Deborah Allen, for her suggestions, assistance, and ever cheerful support.

Additional thanks to Meridian Diagnostics in Cincinnati for their generous offer to make diagnostic kits available for some exercises. We also thank the following instructors for their valuable input on the revision of this manual.

Reviewers

Barbara Beck
Rochester Community and Technical College

Mark Chatfield
West Virginia State College

Kathleen C. Smith
Emory University

Evert Ting
Purdue University Calumet

Robert Walters
James Madison University

Laboratory Safety

To be read by the student before beginning any laboratory work.

1. Do not eat, drink, smoke, or store food in the laboratory. Avoid all finger-to-mouth contact.
2. Never pipette by mouth because of the danger of ingesting microorganisms or toxic chemicals.
3. Wear a laboratory coat while in the laboratory. Remove it before leaving the room and store it in the laboratory until the end of the course.*
4. Wipe down the bench surface with disinfectant before and after each laboratory period.
5. Tie long hair back to prevent it from catching fire in the Bunsen burner or contaminating cultures.
6. Keep the workbench clear of any unnecessary books or other items. Do not work on top of the manual because if spills occur, it cannot be disinfected easily.
7. Be careful with the Bunsen burner. Make sure that paper, alcohol, the gas hose, and your microscope are not close to the flame.
8. All contaminated material and cultures must be placed in the proper containers for autoclaving before disposal or washing.
9. Avoid creating aerosols by gently mixing cultures. Clean off the loop in a sand jar before flaming in the Bunsen burner.
10. If a culture is dropped and broken, notify the instructor. Cover the contaminated area with a paper towel and pour disinfectant over the material. After ten minutes, put the material in a broken glass container to be autoclaved.
11. Carefully follow the techniques of handling cultures as demonstrated by the instructor.
12. When the laboratory is in session, the doors and windows should be shut. A sign should be posted on the door indicating that it is a microbiology laboratory.
13. Be sure you know the location of fire extinguishers, eyewash apparatus, and other safety equipment.
14. Wash your hands with soap and water after any possible contamination and at the end of the laboratory period.
15. If you are immunocompromised for any reason (including pregnancy), it may be wise to consult a physician before taking this class.

---

* Other protective clothing includes closed shoes, gloves (optional), and eye protection.

Laboratory Safety Agreement

To be read by the student before beginning any laboratory work.

1. Do not eat, drink, smoke, or store food in the laboratory. Avoid all finger-to-mouth contact.
2. Never pipette by mouth because of the danger of ingesting microorganisms or toxic chemicals.
3. Wear a laboratory coat while in the laboratory. Remove it before leaving the room and store it in the laboratory until the end of the course.*
4. Wipe down the bench surface with disinfectant before and after each laboratory period.
5. Tie long hair back to prevent it from catching fire in the Bunsen burner or contaminating cultures.
6. Keep the workbench clear of any unnecessary books or other items. Do not work on top of the manual because if spills occur, it cannot be disinfected easily.
7. Be careful with the Bunsen burner. Make sure that paper, alcohol, the gas hose, and your microscope are not close to the flame.
8. All contaminated material and cultures must be placed in the proper containers for autoclaving before disposal or washing.

9. Avoid creating aerosols by gently mixing cultures. Clean off the loop in a sand jar before flaming in the Bunsen burner.
10. If a culture is dropped and broken, notify the instructor. Cover the contaminated area with a paper towel and pour disinfectant over the material. After ten minutes, put the material in a broken glass container to be autoclaved.
11. Carefully follow the techniques of handling cultures as demonstrated by the instructor.
12. When the laboratory is in session, the doors and windows should be shut. A sign should be posted on the door indicating that it is a microbiology laboratory.
13. Be sure you know the location of fire extinguishers, eyewash apparatus, and other safety equipment.
14. Wash your hands with soap and water after any possible contamination and at the end of the laboratory period.
15. If you are immunocompromised for any reason (including pregnancy), it may be wise to consult a physician before taking this class.

---

* Other protective clothing includes closed shoes, gloves (optional), and eye protection.

I have read and understood the laboratory safety rules:

_____	_____
Signature	Date

*I*NTRODUCTION to Microbiology

When you take a microbiology class, you have an opportunity to explore an extremely small biological world that exists unseen in our own ordinary world. Fortunately, we were born after the microscope was perfected so we can see these extremely small organisms.

A few of these many and varied organisms are pathogens (capable of causing disease). Special techniques have been developed to isolate and identify them as well as to control or prevent their growth. The exercises in this manual will emphasize medical applications. The goal is to teach you basic techniques and concepts that will be useful to you now or can be used as a foundation for additional courses. In addition, these exercises are also designed to help you understand basic biological concepts that are the foundation for applications in all fields.

As you study microbiology, it is also important to appreciate the essential contributions of microorganisms as well as their ability to cause disease. Most organisms play indispensable roles in breaking down dead plant and animal material into basic substances that can be used by other growing plants and animals. Photosynthetic bacteria are an important source of the earth's supply of oxygen. Microorganisms also make major contributions in the fields of antibiotic production, food and beverage production as well as food preservation, and more recently, recombinant DNA technology. The principles and techniques demonstrated here can be applied to these fields as well as to medical technology, nursing, or patient care. This course is an introduction to the microbial world, and we hope you will find it useful and interesting.

Note: The use of pathogenic organisms has been avoided whenever possible, and nonpathogens have been used to illustrate the kinds of tests and procedures that are actually carried out in clinical laboratories. In some cases, however, it is difficult to find a substitute and organisms of low pathogenicity are used. These exercises will have an additional safety precaution.

NOTES:

EXERCISE  1

Ubiquity of Microorganisms

Getting Started

Microorganisms are everywhere—in the air, soil, and water; on plant and rock surfaces; and even in such unlikely places as Yellowstone hot springs and Antarctic ice. Millions of microorganisms are also found living with animals—for example, the mouth, the skin, the intestine all support huge populations of bacteria. In fact, the interior of healthy plant and animal tissues is one of the few places free of microorganisms. In this exercise, you will sample material from the surroundings and your body to determine what organisms are present that will grow on laboratory media.

An important point to remember as you try to grow organisms, is that there is no one condition or **medium** that will permit the growth of all microorganisms. The trypticase soy **agar** used in this exercise is a rich medium (a digest of meat and soy products, similar to a beef and vegetable broth) and will support the growth of many diverse organisms, but bacteria growing in a freshwater lake that is very low in organic compounds would find it too rich (similar to a goldfish in vegetable soup). However, organisms that are accustomed to living in our nutrient-rich throat might find the same medium lacking necessary substances they require.

Temperature is also important. Organisms associated with warm-blooded animals usually prefer temperatures close to 37°C, which is approximately the body temperature of most animals. Soil organisms generally prefer a cooler temperature of 30°C. Organisms growing on glaciers would find room temperature (about 25°C) much too warm and would probably grow better in the refrigerator.

Microorganisms also need the correct atmosphere. Many bacteria require oxygen, while other organisms find it extremely toxic and will only grow in the absence of air. Therefore, the organisms you see growing on the plates may be only a small sample of the organisms originally present.

Definitions

Agar. A carbohydrate derived from seaweed used to solidify a liquid medium.

Colony. A visible population of microorganisms growing on a solid medium.

Inoculate. To transfer organisms to a medium to initiate growth.

Media (medium, singular). The substances used to support the growth of microorganisms.

Pathogen. An organism capable of causing disease.

Sterile. The absence of either viable microorganisms or viruses capable of reproduction.

Ubiquity. The existence of something everywhere at the same time.

Objectives

1. To demonstrate that organisms are **ubiquitous.**
2. To demonstrate how organisms are grown on laboratory culture media.

Reference

Nester et al. *Microbiology: A human perspective,* 4th ed., 2004. Chapter 4.

Materials

Per team of two (or each individual, depending on amount of plates available)

Trypticase soy agar (TSA) plates, 2

Sterile swabs as needed

Sterile water (about 1 ml/tube) as needed

Waterproof marking pen or wax pencil

Procedure

First Session

1. Each pair of two students should obtain two petri plates of trypticase soy agar. Notice that the lid of a petri plate fits loosely over the bottom half.

2. Label the plates with your name and date using a wax pencil or waterproof marker. Always label the bottom of the plate because sometimes you may be examining many plates at the same time and it is easy to switch the lids.

3. Divide each plate in quarters with two lines on the back of the petri plate. Label one plate 37°C and the other 25°C (figure 1.1).

4. **Inoculate** the 37°C plate with samples from your body. For example, moisten a sterile swab with sterile water and rub it on your skin and then on one of the quadrants. Try touching your fingers to the agar before and after washing or place a hair on the plate. Try whatever interests you. (Be sure to place all used swabs into an autoclave container or bucket of disinfectant after use.)

5. Inoculate the plate labeled 25°C (room temperature) with samples from the room. It is easier to pick up a sample if the swab is moistened in sterile water first. **Sterile** water is used so that there will be no living organisms in the water to contaminate your results. Try sampling the bottom of your shoe or some dust, or press a coin or other objects lightly on the agar. Be sure to label each quadrant so that you will know what you used as inoculum.

6. Incubate the plates at the temperature written on the plate. Place the plates in the incubator or basket upside down. This is important because it prevents condensation from forming on the lid and dripping on the agar below. The added moisture would permit colonies of bacteria to run together.

Second Session

Handle all plates with colonies as if they were potential **pathogens.** Follow your instructor's directions carefully.

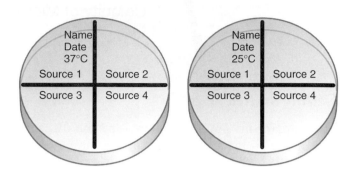

Figure 1.1 Plates labeled on the bottom for ubiquity exercise.

Note: For best results, the plates incubated at 37°C should be observed after 2 days, but the plates at room temperature will be more interesting at about 5–7 days. If possible, place the 37°C plates either in the refrigerator or at room temperature after 2 days so that all the plates can be observed at the same time.

1. Examine the plates you prepared in the first session and record your observations on the report sheet for this exercise. There will be basically two kinds of **colonies:** fungi (molds) and bacteria. Mold colonies are usually large and fluffy, the type found on spoiled bread. Bacterial colonies are usually soft and glistening, and tend to be cream colored or yellow. Compare your colonies with color plates 1 and 2.

2. When describing the colonies include:
   a. relative size as compared to other colonies
   b. shape (round or irregular)
   c. color
   d. surface (shiny or dull)
   e. consistency (dry, moist, or mucoid)
   f. elevation (flat, craterlike, or conical)

3. There may be surprising results. If you pressed your fingers to the agar before and after washing, you may find more organisms on the plate after you washed your hands. The explanation is that your skin has a normal flora (organisms that are always found growing on your skin). When you wash your hands, you wash off the organisms you have picked up from your surroundings as well as a few layers of skin. This exposes more of your normal flora; therefore, you may see different

colonies of bacteria before you wash your hands than afterward. Your flora is important in preventing undesirable organisms from growing on your skin. Hand washing is an excellent method for removing pathogens that are not part of your normal flora.

4. (Optional) If desired, use these plates to practice making simple stains or Gram stains in exercises 4 and 5.

Note: In some labs, plates with molds are opened as little as possible and immediately discarded in an autoclave container to prevent contaminating the lab with mold spores.

5. Follow the instructor's directions for discarding plates. All agar plates are autoclaved before washing or discarding in the municipal garbage system.

NOTES:

EXERCISE

7

Laboratory Report: Ubiquity of Microorganisms

Results

Room Temperature (about 25°C) Plate				
	Plate Quadrant			
	1	2	3	4
Source				
Colony appearance				

37°C Plate				
	Plate Quadrant			
	1	2	3	4
Source				
Colony appearance				

Questions

1. Give three reasons why all the organisms you placed on the TS agar plates might not grow.

2. Why were some agar plates incubated at 37°C and others at room temperature?

3. Why do you invert agar plates when placing them in the incubator?

4. Name one place that might be free of microorganisms.

EXERCISE

Bright-field Light Microscopy, Including History and Working Principles

Getting Started

Microbiology is the study of living organisms too small to be seen with the naked eye. An optical instrument, the microscope, allows you to magnify microbial cells sufficiently for visualization. The objectives of this exercise are to inform you about: (1) some pertinent principles of microscopy; and (2) the practical use, including instruction and care, of the bright-field light microscope.

Historical

Anton van Leeuwenhoek (1632–1723), a Dutch linen draper and haberdasher, recorded the first observations of living microorganisms using a homemade microscope containing a single glass lens (figure 2.1) powerful enough to enable him to see what he described as little "animalcules" (now known as bacteria) in scrapings from his teeth, and larger "animalcules" (now known as protozoa and algae)

Figure 2.1 Model of a van Leeuwenhoek microscope. The original was made in 1673 and could magnify the object being viewed almost 300 times. The object being viewed is brought into focus with the adjusting screws. This replica was made according to the directions given in the *American Biology Teacher* 30:537, 1958. Note its small size. Photograph Courtesy of J.P. Dalmasso

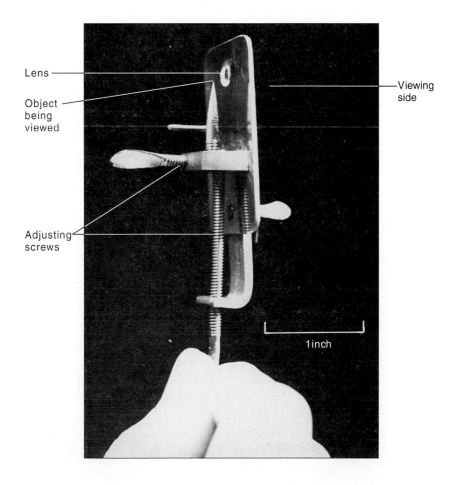

Figure 2.2   Modern bright-field compound microscope. Courtesy of Carl Zeiss, Inc.

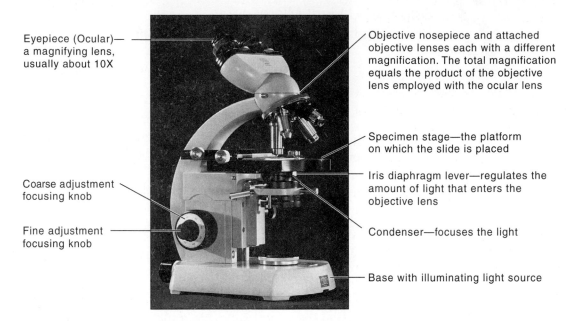

Eyepiece (Ocular)—a magnifying lens, usually about 10X

Objective nosepiece and attached objective lenses each with a different magnification. The total magnification equals the product of the objective lens employed with the ocular lens

Specimen stage—the platform on which the slide is placed

Coarse adjustment focusing knob

Iris diaphragm lever—regulates the amount of light that enters the objective lens

Fine adjustment focusing knob

Condenser—focuses the light

Base with illuminating light source

in droplets of pond water and hay infusions. A single lens microscope such as van Leeuwenhoek's had many disadvantages. Optically, they included production of distortion with increasing magnifying powers and a decrease in focal length (the distance between the specimen when in focus and the tip of the lens). Thus, when using a single lens with an increased magnifying power, van Leeuwenhoek had to practically push his eye into the lens in order to see anything.

Today's microscopes have two lenses, an ocular lens and an objective lens (see figure 2.2). The ocular lens allows comfortable viewing of the specimen from a distance. It also has some magnification capability, usually 10 times (10×) or 20 times (20×). The purpose of the objective lens, which is located near the specimen, is to provide image magnification and image clarity. Most teaching microscopes have three objective lenses with different powers of magnification (usually 10×, 45×, and 100×). Total magnification is obtained by multiplying the magnification of the ocular lens by the magnification of the objective lens. Thus, when using a 10× ocular lens with a 45× objective lens, the total magnification of the specimen image is 450 diameters.

Another giant in the early development of the microscope was a German physicist, Ernst Abbe, who (ca. 1883) developed various microscope im-

provements. One was the addition of a third lens, the **condenser lens**, which is located below the microscope stage (see figure 2.2). By moving this lens up or down, it becomes possible to concentrate (intensify) the light emanating from the light source on the bottom side of the specimen slide. The specimen is located on the top surface of the slide.

He also developed the technique of using lens immersion oil in place of water as a medium for transmission of light rays from the specimen to the lens of the **oil immersion objective.** Oil with a density more akin to the microscope lens than that of water helps to decrease the loss of transmitted light, which, in turn, increases image clarity. Finally, Abbe developed improved microscope objective lenses that were able to reduce both **chromatic** and **spherical lens aberrations.** His objectives include the addition of a concave (glass bent inward like a dish) lens to the basic convex lens (glass bent outward). Such a combination diverges the peripheral rays of light only slightly to form an almost flat image. The earlier simple convex lenses produced distorted image shapes due to spherical lens aberrations and distorted image colors due to chromatic lens aberrations.

Spherical Lens Aberrations   These occur because light rays passing through the edge of a convex lens are bent more than light rays passing through the center. The simplest correction is the placement of a diaphragm below the lens so that only the center of the lens is used (locate **iris diaphragm** in figure 2.2). Such aberrations can also be corrected by grinding the lenses in special ways.

Chromatic Lens Aberrations These occur because light is refracted (bent) as well as dispersed by a lens. The blue components of light are bent more than the red components. Consequently, the blue light, which is bent the most, travels a shorter distance through the lens before converging to form a blue image. The red components, which are bent the least, travel a longer distance before converging to form a red image. When these two images are seen in front view, the central area, in which all the colors are superimposed, maintains a white appearance. The red image, which is larger than the blue image, projects beyond the central area, forming red edges outside of the central white image. Correction of a chromatic aberration is much more difficult than correction of a spherical aberration since dispersion differs in different kinds of glass. Objective lenses free of spherical and chromatic aberrations, known as **apochromatic objectives,** are now available but are also considerably more expensive than **achromatic objectives.**

Some Working Principles of Bright-field Light Microscopy

Subjects for discussion include microscope objectives, magnification and resolution, and illumination.

Microscope Objectives—The Heart of the Microscope

All other parts of the microscope are involved in helping the objective attain a noteworthy image. Such an image is not necessarily the *largest* but the *clearest.* A clear image helps achieve a better understanding of specimen structure. Size alone does not help achieve this end. The ability of the microscope to reveal specimen structure is termed **resolution,** whereas the ability of the microscope to increase specimen size is termed **magnification.**

Resolution or resolving power is also defined as the ability of an objective to distinguish two nearby points as distinct and separate. The maximum resolving power of the human eye when reading is 0.1 mm (100 micrometers). We now know that the maxi-mum resolving power of the light microscope is approximately 0.2 μm or 500× better than the human eye, and that it is dependent on the wavelength (λ) of light used for illumination, and the numerical apertures (NA) of the objective and condenser lens systems. These are related by the equation:

$$\text{resolving power (r)} = \frac{\lambda}{\text{NA}_{obj} + \text{NA}_{cond}}$$

Examining the above equation, we can see that the resolving power can be increased by decreasing the wavelength and by increasing the **numerical aperture.** Blue light affords a better resolving power than red light because its wavelength is considerably shorter. However, because the range of the visible light spectrum is rather narrow, increasing the resolution by decreasing the wavelength is of limited use. Thus, the greatest boost to the resolving power is attained by increasing the numerical aperture of the condenser and objective lens systems.

By definition, the numerical aperture = n sin theta. The **refractive index,** n, refers to the medium employed between the objective lens and the upper slide surface as well as the medium employed between the lower slide surface and the condenser lens. With the low and high power objectives the medium is air, which has a refractive index of 1, whereas with the oil immersion objective the medium is oil, which has a refractive index of 1.25 or 1.56. Sin theta is the maximum angle formed by the light rays coming from the condenser and passing through the specimen into the front lens of the objective.

Ideally, the numerical aperture of the condenser should be as large as the numerical aperture of the objective, or the latter is reduced, resulting in reduced resolution. Practically, however, the condenser numerical aperture is somewhat less because the condenser iris has to be closed partially in order to avoid glare. It is also important to remember that the numerical aperture of the oil immersion objective depends upon the use of a dispersing medium with a refractive index greater than that of air ($n = 1$). This is achieved by using oil, which must be in contact with both the condenser lens (below the slide) and the objective lens (above the slide).

Note: Oil should not be placed on the surface of the condenser lens unless your microscope contains an oil immersion type condenser lens and your instructor authorizes its use.

When immersion oil is used on only one side of the slide, the maximum numerical aperture of the oil immersion objective is 1.25—almost the same as the refractive index of air.

Microscopes for bacteriological use are usually equipped with three objectives: 16 mm low power (10×), 4 mm high dry power (40 to 45×), and 1.8 mm oil immersion (100×). The desired objective is rotated into place by means of a revolving nosepiece (see figure 2.2). The millimeter number (16, 4, 1.8) refers to the **focal length** of each objective. By definition, the focal length is the distance from the principal point of focus of the objective lens to the principal point of focus of the specimen. Practically speaking, one can say that the shorter the focal length of the objective, the shorter the **working distance** (that is, the distance between the lens and the specimen) and the larger the opening of the condenser iris diaphragm required for proper illumination (figure 2.3).

The power of magnification of the three objectives is indicated by the designation 10×, 45×, and 96× inscribed on their sides (note that these values may vary somewhat depending upon the particular manufacturer's specifications). The total magnification is obtained by multiplying the magnification of the objective by the magnification of the ocular eyepiece. For example, the total magnification obtained with a 4 mm objective (45×) and a 10× oc-

ular eyepiece is 45 × 10 = 450 diameters. The highest magnification is obtained with the oil immersion objective. The bottom tip lens of this objective is very small and admits little light, which is why the iris diaphragm of the condenser must be wide open and the light conserved by means of immersion oil. The oil fills the space between the object and the objective so light is not lost (see figure 2.4 for visual explanation).

Microscope Illumination

Proper illumination is an integral part of microscopy. We cannot expect a first-class microscope to produce the best results when using a second-class illuminator. However, a first-class illuminator improves a second-class microscope almost beyond the imagination. A student microscope with only a mirror (no condenser) for illumination can be operated effectively by employing light from a gooseneck lamp containing a frosted or opalescent bulb. Illuminators consisting of a sheet of ground glass in front of a clear bulb are available but they offer no advantage over a gooseneck lamp. Microscope mirrors are flat on one side and concave on the other. In the absence of a condenser, the concave side of the mirror should be used. Conversely, with a condenser the flat side of the mirror should be used since condensers accept only parallel rays of light and focus them on the slide.

Figure 2.3 Relationship between working distance of objective lens and the diameter of the opening of the condenser iris diaphragm. The larger the working distance, the smaller the opening of the iris diaphragm.

Figure 2.4 This diagram shows that light refracts (bends) more when it passes through air (refractive index $n = 1$) than when it passes through oil ($n = 1.6$). Thus, by first passing the light from the light source through oil, light energy is conserved. This conservation in light energy helps to increase the resolving power of the oil immersion objective, which also has a refractive index greater than 1 ($n = 1.25$ to 1.35).

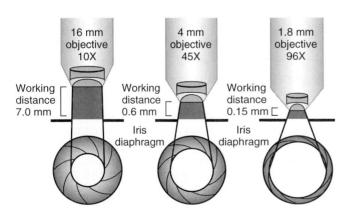

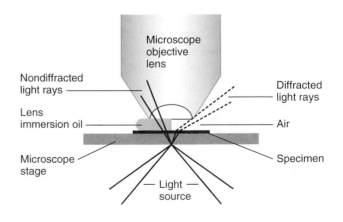

Condensers with two or more lenses are necessary for obtaining the desired numerical aperture. The Abbe condenser, which has a numerical aperture of 1.25, is most frequently used. The amount of light entering the objective is regulated by opening and closing the iris diaphragm located between the condenser and the light source (see figure 2.2). When the oil immersion objective is used, the iris diaphragm is opened farther than when the high dry or low power objectives are used. Focusing the light is controlled by raising or lowering the condenser by means of a condenser knob.

The mirror, condenser, and objective and ocular lenses must be kept clean to obtain optimal viewing. The ocular lenses are highly susceptible to etching from acids present in body sweat and should be cleaned after each use. (See step 6 below.)

Precautions for Proper Use and Care of the Microscope

Your microscope is a precision instrument with delicate moving parts and lenses. Instruction for proper use and care is as follows:

1. Use both hands to transport the microscope. Keep upright. If inverted, oculars may fall out.
2. Do not touch lenses with your hands. Use lens paper instead. Use of other cleaning materials such as handkerchiefs and Kleenex tissues is discouraged because they may scratch the lens.
3. Do not force any of the various microscope adjustment knobs. If you experience problems making adjustments, consult your instructor.
4. Do not remove objective or ocular lenses for cleaning, or exchange them with different microscopes.
5. For routine cleaning of the oil immersion objective lens, it is necessary only to wipe off excess oil with a piece of dry lens paper. Any special cleaning should be done under the guidance of the instructor.
6. Before storing the microscope, make certain that the ocular lens is also clean. Frequently, sweat deposits from your eyes, which are acidic, can etch the glass. The presence of other foreign particles can be determined by rotating the ocular lens manually as you look through the microscope. The presence of a pattern that rotates is evidence of dirt. Clean the upper and lower surfaces of the ocular with lens paper moistened with a drop of distilled water. If dirt persists, consult your instructor. Any dirt remaining after cleaning with a suitable solvent indicates either a scratched lens surface or the presence of dirt on the inside surface of the lens.
7. A blast of air from an air syringe may be effective in removing any remaining dust particles from the lenses.

Definitions

Achromatic objective. A microscope objective lens in which the light emerging from the lens forms images practically free from prismatic colors.

Apochromatic objective. A microscope objective lens in which the light emerging from the lens forms images practically free from both spherical and chromatic aberrations.

Bright-field light microscopy. A form of microscopy in which the field is bright and the specimen appears opaque.

Chromatic lens aberration. A distortion in the lens caused by the different refrangibilities of the colors in the visible spectrum.

Compound microscope. A microscope with more than one lens.

Condenser. A structure located below the microscope stage that contains a lens and iris diaphragm. It can be raised or lowered, and is used for concentrating and focusing light from the illumination source on the specimen.

Focal length. The distance from the principal point of a lens to the principal point of focus of the specimen.

Iris diaphragm. An adjustable opening that can be used to regulate the aperture of a lens.

Magnification. The ability of a microscope to increase specimen size.

Numerical aperture. A quantity that indicates the resolving power of an objective. It is numerically equal to the product of the index of refraction of the medium in front of the objective lens (n) and the sine of the angle that the most oblique light ray entering the objective lens makes with the optical axis.

Parfocal.   Having a set of objectives so mounted on the microscope that they can be interchanged without having to appreciably vary the focus.

Refractive index.   The ratio of the velocity of light in the first of two media to its velocity in the second medium as it passes from one medium into another medium with a different index of refraction.

Resolution.   The smallest separation which two structural forms, *e.g.*, two adjacent cilia, must have in order to be distinguished optically as separate cilia.

Simple microscope.   A microscope with only one lens.

Spherical lens aberration.   An aberration caused by the spherical form of a lens that gives different focal lengths for central and marginal light rays.

Wet mount.   A microscope slide preparation in which the specimen is immersed in a drop of liquid and covered with a coverslip.

Working distance.   The distance between the tip of the objective lens when in focus and the slide specimen.

Objectives

1. Introduction of historical information on microscopy development from van Leeuwenhoek's single lens light microscope to the compound light microscope of today.
2. Introduction of some major principles of light microscopy, including proper use and care of the microscope.
3. To teach you how to use the microscope and become comfortable with it.

References

Dobell, C. *Anton van Leeuwenhoek and his "little animals."* New York: Dover Publications, Inc., 1960.

Gerhardt, P.; Murray, R. G. E.; Costillo, R. N.; Nester, E. W.; Wood, W. A.; Krieg, N. R.; and Phillips, G. B., eds. *Manual of methods for general bacteriology.* Washington, D.C.: American Society for Microbiology, 1981. Contains three excellent chapters on principles of light microscopy.

Gray, P., ed. *Encyclopedia of microscopy and microtechnique.* New York: Van Nostrand-Reinhold, 1973.

Lechevalier, Hubert A., and Solotorovsky, Morris. *Three centuries of microbiology.* New York: McGraw-Hill, 1965. Excellent history of microbiology showing how scientists who made these discoveries were often influenced by other developments in their lives.

Nester et al. *Microbiology: A human perspective*, 4th ed., 2004. Chapter 3. Other types of light microscopy are also discussed in this chapter.

Materials

Cake of baker's yeast (sufficient for entire class)

Tube containing 10 ml distilled water (one per student)

Plastic dropper (one per student)

Prepared stained slides of various bacterial forms (coccus, rod, spiral), sufficient for entire class

Procedure

1. Place the microscope on a clear space on your desk, and identify the different parts with the aid of figure 2.2.
2. Before using it be sure to read the Getting Started section titled "Precautions for Proper Use and Care of the Microscope."
3. Sample preparation (**wet mount**). Prepare a yeast cell suspension by adding to water in a test tube just enough yeast to cause visible clouding (approximately 1 loopful per 10 ml of water). Remove a small amount of the suspension with a plastic dropper and carefully place a drop on the surface of a clean slide. Cover the drop with a clean coverslip. Discard dropper as directed by instructor.

4. Place the wet mount in the mechanical slide holder of the microscope stage with the coverslip side up. Center the coverslip with the mechanical stage control over the stage aperture.

5. Practice focusing and adjusting light intensity when using the low and high power objectives. Rotate the low power objective (10× if available) in position. To focus the objective, you must decrease the distance between the objective lens and the slide. This is done by means of the focusing knobs on the side of the microscope (see figure 2.2). Movement of these knobs on some microscopes causes the objective lens to move up and down in relation to the stage; in other microscopes, the stage moves up and down in relation to the objective. For initial, so-called coarse focusing, the larger adjustment knob is used. For final, so-called fine focusing, the smaller adjustment knob is used. With the large knob, bring the yeast cells into coarse focus. Then complete the focusing process with the fine adjustment knob. Remember that the objective lens should never touch the surface of the slide or coverslip. This precaution helps prevent scratching of the objective lens and (or) cracking of the slide.

   Adjust the light intensity to obtain optimal image detail by raising or lowering the condenser and by opening or closing the iris diaphragm. For best results, keep the condenser lens at the highest level possible because at lower levels the resolving power is reduced. After examining and drawing a few yeast cells under low power, proceed to the high dry objective by rotating the nosepiece until it clicks into place. If the microscope is parfocal, the yeast cells will already have been brought into approximate focus with the low power so that only fine focusing will be required. Once again, adjust the iris diaphragm and condenser for optimal lighting. If the microscope is not parfocal, it will be necessary, depending on the type of microscope, either to lower the body tube or to raise the stage with the coarse adjustment knob until it is about $^1/_{16}$ inch from the coverslip surface. Repeat these steps to focus the high power objective. Note the increased size of the yeast cells and the decreased number of cells present per microscopic field. Draw a few representative cells (see color plate 6 and Laboratory Report).

6. Focusing with the oil immersion objective. First rotate the high dry objective to one side so that a small drop of lens immersion oil may be placed on the central surface of the coverslip. Slowly rotate the oil immersion objective into place. The objective lens should be in the oil but should not contact the coverslip. Next bring the specimen into coarse focus very slowly with the coarse adjustment knob, and then into sharp focus with the fine adjustment knob. The yeast cells will come into view and go out of view quickly because the depth of focus of the oil immersion objective is very short. Refocus when necessary. Draw a few cells.

7. Examine the prepared stained bacteria slides with the oil immersion objective. (See exercise 4, Procedure, "Simple Stain" step 12 for information on how to prepare and focus stained slides with the oil immersion objective.) Once again, if your microscope is parfocal, first focus the slide with the lower power objective before using the oil immersion objective. Draw a few cells of each bacterial form. Compare the shapes of these cells with those in color plates 3–5.

8. When you finish this procedure, wipe the excess oil from the oil immersion objective with lens paper, and if necessary clean the ocular (see "Precautions for Proper Use and Care of the Microscope"). Next return the objective to the low power setting, and if your microscope has an adjustable body tube, lower (rack down) it before returning the microscope to the microscope cabinet.

NOTES:

Name _____ Date _____ Section _____

EXERCISE

Laboratory Report: Bright-field Light Microscopy, Including History and Working Principles

Results

1. Draw a few yeast cells from each magnification. Include any interesting structural changes evident at the three magnifications.

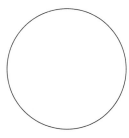 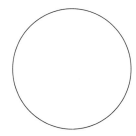 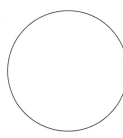

Magnification: _____   _____   _____

Objective: _____   _____   _____

2. Examination of prepared bacteria slides. Examine with the oil immersion objective and draw a few cells of each morphological form.

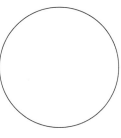 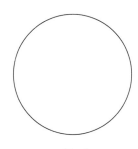 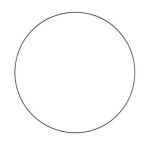

Coccus   Rod   Spiral

3. Answer the following questions about your microscope:
   a. What is the magnification and numerical aperture (NA) stamped on each objective of your microscope?

Objective Magnification	Numerical Aperture
_____	_____
_____	_____
_____	_____
_____	_____

4. What is the magnification stamped on the oculars? _____
5. Calculate the total magnification of the objective/ocular combination with:

The lowest power objective: _____

The highest power objective: _____

Questions

1. Discuss the advantages of a modern compound microscope (figure 2.2) over an early microscope (figure 2.1).

   _____

   _____

   _____

2. Why must the distance from slide to objective increase rather than decrease when coarse focusing with the high dry and oil immersion objectives?

   _____

3. How does increasing the magnification affect the resolving power?

   _____

4. How does lens immersion oil help to increase the resolving power of the oil immersion objective?

   _____

   _____

5. How can you determine that the ocular and objective lenses are free of sweat, oil, and dust contaminants?

   _____

   _____

6. What are the functions of the substage condenser?

   _____

   _____

7. What is meant by the term "parfocal"? Does it apply to your microscope?

   _____

True-False Questions

Mark the statements below true (T) or false (F).

1. Van Leeuwenhoek's microscope was corrected for spherical but not chromatic aberrations. _____

2. Spherical lens aberrations are easier to correct than chromatic lens aberrations. _____

3. The objective NA is more important than the condenser NA for increasing resolving power. _____

4. The working distance is the distance from the tip of the objective to the tip of the condenser lens. _____

5. Excess oil on the oil immersion objective can safely be removed with lens paper containing a drop of solvent. _____

NOTES:

EXERCISE 3

Microscopic (Bright-field and Dark-field) Determination of Cell Motility, Form, and Viability Using Wet Mount and Hanging Drop Preparations

Getting Started

Although bacterial cell motility is usually determined by the semisolid agar stab inoculation method, it is sometimes determined by direct microscopic examination. Microscopic examination allows for the determination of cell form, for example, their general shape (round or coccus, elongate or rod, etc.); and their arrangement, for example, how the cells adhere and attach to one another (as filaments, tetrads, etc.). It is also sometimes possible to determine cell viability using either bright-field microscopy and a **vital stain** or dark-field microscopy without a stain. With dark-field microscopy, living cells appear bright and dead cells appear dull. With bright-field microscopy and methylene blue stain, living cells appear colorless, whereas dead cells appear blue. The dead cells are unable to enzymatically reduce methylene blue to the colorless form.

For all of the above methods, a **wet mount slide** or a **hanging drop slide** cell preparation is used. Wet mounts are easier to prepare but dry out more rapidly due to contact between the coverslip and air on all four sides. The drying out process can sometimes create false motility positives. Drying out can be reduced by ringing the coverslip edges with petroleum jelly. Other disadvantages are the inability at times to see the microorganism because it is not sufficiently different in refractive index from the suspending fluid (this can sometimes be resolved by reducing the light intensity). It is not particularly useful for observing thick preparations such as hay infusions.

In this exercise, bright-field microscopy is used with wet mounts to observe bacterial motility and form. In observing bacterial motility, it is important to distinguish true motility from "Brownian movement," a form of movement caused by molecules in the liquid striking a solid object, in this instance the bacterial cell, causing it to vibrate back and forth. If the bacterial cell is truly motile, you will observe its directional movement from point A to point B, providing the cells are not in the **resting stage** of the growth curve.

Measurement of cell viability with methylene blue may also be skewed. When resting stage cells are used (Kleyn et al., 1962) they, although viable, are often unable to reduce the dye to a colorless form. Thus, it is preferable to observe cells from the early logarithmic stage of the growth curve (see figure 10.1). The cells of choice—yeast—are sufficiently large for ease of observation with bright-field microscopy when using the high dry objective. Unstained cells from the same stage of the growth curve will also be observed for viability by using dark-field microscopy. Thus, you will be able to compare viability results for the two methods with one another. Hopefully they will vary no more than 10%—one accepted standard of error for biological material.

Definitions

Dark-field microscopy. A form of microscopy in which the specimen is brightly illuminated on a dark background.

Depression slide. A microscope slide with a circular depression in its center.

Hanging drop slide. A microscopic specimen observation technique in which the specimen hangs suspended from an inverted coverslip mounted on a depression slide.

Resting stage. The stage of the growth curve in which cells are metabolically inactive.

Star diaphragm. A metal diaphragm used for dark-field microscopy. Its opaque center deflects the light rays that converge on the objective so that only the oblique rays strike the specimen. The net result is a dark-colored microscope field with a brightly colored specimen.

Vital stain. A stain able to differentiate living from dead cells, e.g., methylene blue is colorless when reduced in the presence of hydrogen, while remaining blue in its absence.

Wet mount slide. A microscopic specimen observation technique in which a drop containing the specimen is placed on the surface of a clean slide, followed by the addition of a coverslip over the drop.

Objectives

1. To become familiar with the advantages and limitations of wet mount and hanging drop preparations for observing living cell material. This will be achieved both by reading and direct experience using living bacteria and yeast cultures as specimen material.
2. To learn how to use dark-field microscopy to observe living cells.

References

Kleyn, J.; Mildner, R.; and Riggs, W. 1962. Yeast viability as determined by methylene blue staining. *Brewers Digest* 37 (6):42–46.

Nester et al. Microbiology: A human perspective, 4th ed., 2004. Chapter 3 and Chapter 4.

Materials

Cultures

12–18 hour nutrient broth cultures of *Staphylococcus epidermidis*, and *Spirillum volutans* showing visible clouding

12–18 hour nutrient broth cultures of *Bacillus cereus* and *Pseudomonas aeruginosa* showing visible clouding

A yeast suspension previously prepared by suspending sufficient baker's yeast in a tube of glucose yeast fermentation broth to produce visible clouding, followed by 6–8 hour incubation at 25°C

A hanging drop depression slide

Vaseline and toothpicks

Pasteur pipets

Dropper bottle with acidified methylene blue

A star diaphragm for dark-field microscopy (figure 3.1)

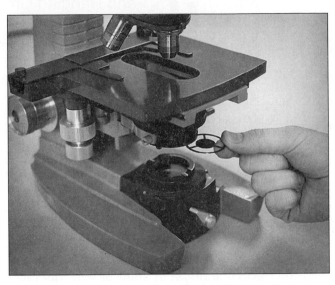

Figure 3.1 Conversion of a bright-field light microscope into a dark-field microscope by inserting a star diaphragm into the filter holder located below the condenser lens.
Courtesy of Dr. Harold J. Benson

Procedure

Wet Mounts for Study of Bacterial Form and Motility

1. Prepare six clean microscope slides and seven clean coverslips by washing them in a mild detergent solution, rinsing with distilled water, and then drying them with a clean towel. Examine visually for clarity.
2. Suspend your broth culture of *S. epidermidis* by gentle tapping on the outside of the culture tube. Hold the tube firmly between thumb and index finger and tap near the bottom of the test tube with your finger until the contents mix.
3. Remove the test tube cover and with a Pasteur pipet, finger pipette approx. 0.1 ml of the broth culture.
4. Transfer a drop of this suspension to the surface of a slide.

   Note: The drop must be of suitable size; if it is too small, it will not fill the space between the coverslip and the slide; if it is too large, some of the drop will pass outside the coverslip, which could smear the front lens of the microscope objective. If such occurs, prepare a fresh wet mount.

   Discard the Pasteur pipet in the designated container.

5. Grasp a clean coverslip on two edges and place it carefully over the surface of the droplet.
6. Insert the wet mount on the stage of your microscope and examine for cell motility and form with the oil immersion objective. Make certain you can distinguish true motility from Brownian movement. Prepare a drawing of some of the cells and record your findings in the Laboratory Report.
7. Discard the slide in the designated container for autoclaving.
8. Repeat the above procedure with S. *volutans*, B. *cereus*, and P. *aeruginosa* (for representative cell shapes see color plates 3–5).

Use of Hanging Drop Slides for Study of Bacterial Form and Motility

1. Prepare a clean depression drop slide and coverslip.
2. With a toothpick, spread a thin ring of Vaseline approximately ¼ inch outside the **depression slide** concavity (figure 3.2a).
3. Using your suspended B. *cereus* broth culture and a wire loop, transfer 2 loopfuls to the central surface of a coverslip (see figure 3.2b).
4. Invert the depression slide and center the depression over the droplet on the coverslip. Make contact and press lightly, forming a seal between the Vaseline ring and coverslip (see figure 3.2c).
5. Quickly turn over the depression slide so as not to disrupt the culture droplet.

   Note: If done correctly, the droplet will remain suspended and will not come in contact with the well bottom.

6. Place the slide on the stage of your microscope and first focus the *edge* of the droplet with your low power objective. You may also need to reduce the light to achieve proper contrast. Due to capillary action, most microorganisms gather at the edge. When in focus, the edge will appear as a light line against a dark background.
7. In order to see individual bacterial cells, you will need to use the oil immersion objective. Add a drop of lens immersion oil to the coverslip, and if parfocal, shift to the oil

immersion objective. Once again, light adjustment becomes necessary. You should now be able to observe individual bacteria, their form, and motility. If not, ask your instructor for help.
8. Draw some of the cells and record their motility and other findings in the Laboratory Report. Discard slide in the designated waste glass container.

Use of Dark-field Microscopy to Determine Yeast Cell Viability

1. Insert the **star diaphragm** into the filter holder located below the microscope condenser (see figure 3.1).

   Note: Make certain that it is accurately centered.
2. Suspend the baker's yeast preparation and prepare a wet mount. Transfer the wet mount to the microscope stage.
3. Examine the wet mount with the low power objective. Keep the iris diaphragm wide open in order to admit as much light as possible.
4. Adjust the condenser focus to the position where the best dark-field effect is obtained. See

---

Figure 3.2 (a-c) Preparation of a hanging drop slide.

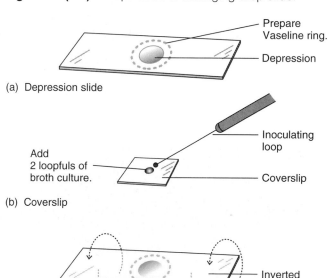

(a) Depression slide

Prepare Vaseline ring.

Depression

Add 2 loopfuls of broth culture.

Inoculating loop

Coverslip

(b) Coverslip

Inverted slide

(c) Pressing of slide against cover glass

color plate 6 for examples of yeast photographed with bright-field and dark-field microscopy.

5. Examine the wet mount with the high dry objective.

   Note: Dark-field microscopy may or may not be possible at this magnification depending upon how well the oblique light rays pass through the objective lens.

6. Determine the percent of viable yeast cells. To do so, count a total number of approximately 100 or more cells and also the number of dull-looking cells (dead cells) within this total. With this information, you can calculate the percent of viable yeast cells.

7. In the Laboratory Report, prepare drawings of representative cells and show your calculations for determining the percent of viable cells.

Use of a Vital Stain, Methylene Blue, to Determine Yeast Viability

1. From a dropping bottle, transfer a small drop of methylene blue to the surface of a clean slide.

2. With a Pasteur pipet, add a small drop of the baker's yeast suspension. Carefully place a clean coverslip over the surface of the droplet.

3. Observe the wet mount with bright-field microscopy using the low and high dry microscope objectives.

4. For the Laboratory Report, prepare drawings of representative cells and show your calculations for determining the percent of viable yeast cells. In this instance, dead cells stain blue and viable cells remain colorless.

EXERCISE

Laboratory Report: Microscopic (Bright-field and Dark-field) Determination of Cell Motility, Form, and Viability Using Wet Mount and Hanging Drop Preparations

Results

1. Wet mounts for study of bacterial form and motility
   Drawings of representative cells showing their relative sizes, shapes, and arrangements. Record magnification (×) and motility (+ or −).

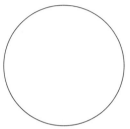 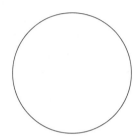  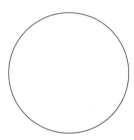

S. epidermidis   *S. volutans*   *B. cereus*   *P. aeruginosa*

_____ ×   _____ ×   _____ ×   _____ ×

_____ motility   _____ motility   _____ motility   _____ motility

2. Hanging drop slide (*B. cereus*)
   Make observations similar to those above and indicate any differences from the *B. cereus* wet mount observations.

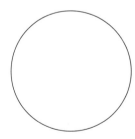

Differences: _____

_____

_____

B. cereus

_____ ×

_____ motility

3. Dark-field microscopy of baker's yeast
   Drawings of cells showing their size, shape, and arrangement, as well as the visual appearance of living and dead cells. Record magnifications used.

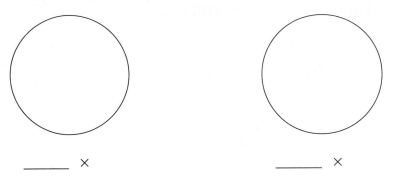

   _____ ×   _____ ×

   Show your calculations for determining the percent of viable cells.

4. Bright-field microscopy of baker's yeast stained with methylene blue
   Make the same kind of observations as in number 3. Record magnifications used.

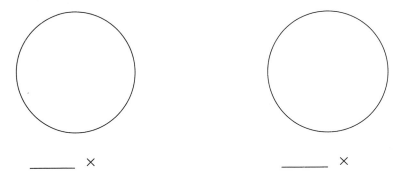

   _____ ×   _____ ×

   Show your calculations for determining the percent of viable cells.

Discuss your yeast cell viability results by the two methods on page 26. If a wide viability variance (> 10%) exists between the two methods, what other method might you use to prove which method is more accurate? You may wish to consult your text (chapter 3) or lab manual (exercise 8) for help in constructing a reasonable answer.

Questions

1. What advantages are there in determining cell motility microscopically rather than with a stab culture?

2. What advantages does a hanging drop preparation have over a wet mount preparation? Disadvantages?

3. How did you obtain optimal results with dark-field microscopy?

4. Why is it difficult to employ the oil immersion objective for dark-field microscopy?

5. What might be a reason for employing an actively multiplying culture when examining viability microscopically?

6. In addition to determining cell viability, what other useful morphological determination can sometimes be made with dark-field microscopy? Consult your text.

7. What difficulties might there be in attempting to determine the viability of bacterial cells with stains such as methylene blue? This will no doubt require some investigation of the literature. A possible clue lies in the prokaryotic makeup of bacteria. Yeasts, on the other hand, are eukaryotic cells.

*I*NTRODUCTION to Staining of Microorganisms

Bacteria are difficult to observe in a broth or wet mount because there is very little contrast between them and the liquid in which they are suspended. This problem is solved by staining bacteria with dyes. Although staining kills bacteria so their motility cannot be observed, the stained organisms contrast with the surrounding background and are much easier to see. The determination of the shape, size, and arrangement of the cells after dividing are all useful in the initial steps in identifying an organism. These can be demonstrated best by making a smear on a glass slide from the clinical material, a broth culture, or a colony from a plate, then staining the smear with a suitable dye. Examining a stained preparation is one of the first steps in identifying an organism.

Staining procedures used here can be classified into two types: the simple stain and the multiple stain. In the **simple stain,** a single stain such as methylene blue or crystal violet is used to dye the bacteria. The shape and the grouping of the organisms can be determined, but all organisms (for the most part) are stained the same color. Another kind of simple stain is the negative stain. In this procedure, the organisms are mixed with a dye and permitted to dry. When they are observed, the organisms are clear against a dark background.

The multiple stain involves more than one stain. The best known example is the Gram stain, which is widely used. After staining, some organisms appear purple and others pink, depending on the structure of their cell wall.

Multiple stains are frequently known as **differential stains** because they are used to visualize special structures of bacteria. In contrast with eukaryotic organisms, prokaryotic organisms have relatively few morphological differences. Several of these structures such as endospores, capsules, acid-fast cell walls, storage bodies, and flagella can be seen with special stains. In the next two exercises, you will have an opportunity to stain bacteria with a variety of staining procedures and observe these structures.

NOTES:

EXERCISE 4

Simple Stains: Positive and Negative Stains

Getting Started

Two kinds of single stains will be done in this exercise: the simple stain and the negative stain. Microbiologists most frequently stain organisms with the Gram stain, but in this exercise a simple stain will be used to give you practice staining and observing bacteria before doing the more complicated multiple, or differential, stains.

After you have stained your bacterial **smears,** you can examine them with the oil immersion lens, which will allow you to distinguish the morphology of different organisms. The typical bacteria you will see are about 0.5–1.0 **micrometer** (μm) in width to about 2–7 μm long and are usually rods, cocci, or spiral-shaped. Sometimes rods are referred to as bacilli, but since that term is also a genus name (*Bacillus*) for a particular organism, the term rod is preferred.

Another kind of simple stain is the **negative stain.** Although it is not used very often, it is advantageous in some situations. Organisms are mixed in a drop of nigrosin or India ink on a glass slide. After drying, the organisms can then be observed under the microscope as clear areas in a black background. This technique is sometimes used to observe capsules or **inclusion bodies.** It also prevents eyestrain when many fields must be scanned. The dye tends to shrink away from the organisms, causing cells to appear larger than they really are.

In both of these simple stains, you will be able to determine the shape of the bacteria and the characteristic grouping after cell division (as you did in the wet mounts). Some organisms tend to stick together after dividing and form chains or irregular clumps. Others are usually observed as individual cells. However, this particular characteristic depends somewhat on how the organisms are grown. *Streptococcus* form long, fragile chains in broth, but if they grow in a colony on a plate, it is sometimes difficult to make a smear with these chains intact.

Definitions

Differential stain.   A procedure that stains specific morphological structures—usually a multiple stain.

Inclusion bodies.   Granules of storage material such as sulfur that accumulate within some bacterial cells.

Micrometer.   (abbreviated μm) The metric unit used to measure bacteria. It is 10^{-6} m (meter) and 10^{-3} mm (millimeter).

Negative stain.   A simple stain in which the organisms appear clear against a dark background.

Parfocal.   If one objective lens of a microscope is in focus, all lenses will be in focus when used.

Simple stain.   A procedure for staining bacteria consisting of a single stain.

Smear.   A dried mixture of bacteria and water (or broth) on a glass slide in preparation for staining.

Objectives

1. Learn to prepare and stain a bacterial smear using a simple stain.
2. Observe stained organisms under the oil immersion lens.
3. Prepare and observe a negative stain.
4. Observe the various morphologies and arrangements of bacteria in stained preparations.

References

Gerhardt, Philip, ed. *Manual for general and molecular bacteriology.* Washington, D.C.: American Society for Microbiology, 1994.

Nester et al. *Microbiology: A human perspective,* 4th ed., 2004. Chapter 3, Section 3.2.

Materials

Cultures

> *Bacillus subtilis* or *B. cereus*
>
> *Staphylococcus epidermidis*
>
> *Enterococcus faecalis*
>
> *Micrococcus luteus*

Staining bottles with:

> crystal violet
>
> methylene blue
>
> safranin

Glass slides

Wax pencils or waterproof marking pen

Tap water in small dropper bottle (optional)

Inoculating loop

Alcohol sand bottle (a small screw cap bottle half full of sand and about three-quarters full of 95% alcohol; figure 4.1)

Procedure

Simple Stain

1. Clean a glass slide by rubbing it with slightly moistened cleansing powder such as Boraxo or Bon Ami. Rinse well and dry with a paper towel. Even new slides should be washed because sometimes they are covered with a protective coating.
2. Draw two or three circles with a waterproof pen or wax pencil on the underside of the slide. If

Figure 4.1 Alcohol bottle and inoculating loop.

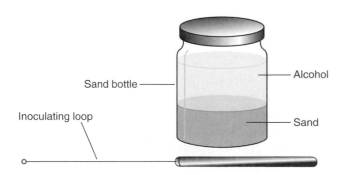

Sand bottle

Inoculating loop

Alcohol

Sand

Figure 4.2 Slide with three drops of water. Three different bacteria can be stained on one slide.

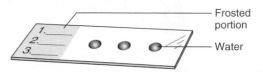

Frosted portion

Water

the slide has a frosted portion, you can also write on it with a pencil. This is useful because it is easy to forget the order in which you placed the organisms on the slide and you can list them, for instance, from left to right (figure 4.2).

3. Add a drop of water to the slide on top of each of the circles. Use your loop to transfer tap water or use water from a dropper bottle. This water does not need to be sterile. Although there are some organisms (nonpathogens) in municipal water systems, there are too few to be seen.

   If you are preparing a smear from a broth culture as you will do in the future, add only the broth to the slide. Broth cultures are relatively dilute, so no additional water is added.

4. Sterilize a loop by holding it at an angle in the flame of the Bunsen burner. Heat the entire wire red hot, but avoid putting your hand directly over the flame or heating the handle itself (figure 4.3).

5. Hold the loop a few seconds to cool it, then remove a small amount of a bacterial culture and suspend it in one of the drops of water on the slide (see figure 4.3). Continue to mix in bacteria until the drop becomes slightly turbid (cloudy). If your preparation is too thick, it will stain unevenly and if it is too thin you will have a difficult time finding organisms under the microscope. In the beginning, it may be better to err on the side of having a slightly too turbid preparation—at least you will be able to see organisms and you will learn from experience how dense to make the suspension.

6. Heat the loop red hot. It is important to burn off the remaining organisms so that you will not contaminate your bench top. If you rest your loop on the side of your Bunsen burner, it can cool without burning anything on the bench.

   Sometimes the cell material remaining on the loop spatters when heated. To prevent

Figure 4.3 Preparation of a bacterial smear.

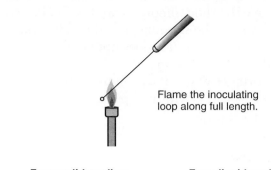

Flame the inoculating loop along full length.

From solid medium   **From liquid medium**

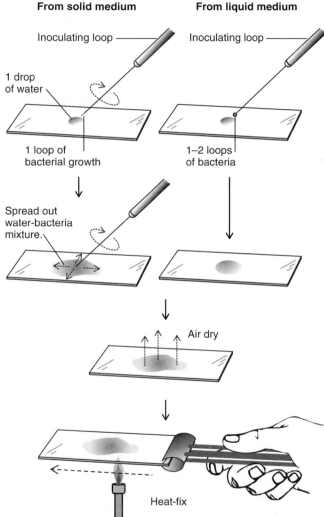

Inoculating loop

Inoculating loop

1 drop of water

1 loop of bacterial growth

1–2 loops of bacteria

Spread out water-bacteria mixture.

Air dry

Heat-fix

this, some laboratories remove bacterial cell material from the loop by dipping the loop in a bottle of sand covered with alcohol. Then the loop is heated red hot in the Bunsen burner.

7. Permit the slide to dry. Do not heat it in any way to hasten the process, since the cells will

become distorted. Place the slide off to the side of the bench so that you can proceed with other work.

8. When the slide is dry (in about 5–10 minutes), heat-fix the organisms to the slide by quickly passing it through a Bunsen burner flame two or three times so that the bottom of the slide is barely warm. This step causes the cells to adhere to the glass so they will not wash off in the staining process (figure 4.4).

9. Place the slide on a staining loop over a sink or pan. Alternatively, hold the slide over the sink

Figure 4.4 (a) Staining, (b) washing, and (c) blotting a simple stain. From John P. Harley and Lansing M. Prescott, *Laboratory Exercises in Microbiology,* 5th ed. Copyright © 2002 The McGraw-Hill Companies. All Rights Reserved. Reprinted by permission.

Simple Staining Procedure

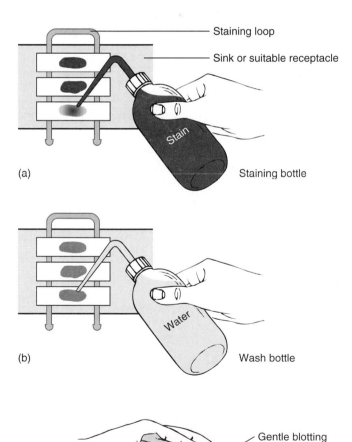

Staining loop

Sink or suitable receptacle

Stain

Staining bottle

(a)

Water

Wash bottle

(b)

Gentle blotting

(c)

with a forceps or clothespin. Cover the specimen with a stain of your choice—crystal violet is probably the easiest to see (figure 4.4).

10. After about 20 seconds, pour off the stain and rinse with tap water (figure 4.4).

11. Carefully blot the smear dry with a paper towel. Do not rub the slide from side to side as that will remove the organisms. Be sure the slide is completely dry (figure 4.4).

12. Observe the slide under the microscope. Since you are looking at bacteria, you must use the oil immersion lens in order to see them. One method is to focus the slide on low power, then cover the smear with immersion oil and move the immersion lens into place. If your microscope is **parfocal,** it should be very close to being in focus. Note that no coverslip is used when looking at stained organisms.

Another method for focusing the oil immersion lens is to put oil on the smear, and then while looking at the microscope from the side very carefully raise the stage (or lower the lens, depending on your microscope) until the immersion lens is just barely touching the slide. Then when looking through the microscope, very slowly back the lens off the slide until it is in focus. Never move the immersion lens toward the slide while looking through the microscope. You may hit the slide with the lens and damage the lens. When you have a particularly thin smear, it is sometimes helpful to put a mark on the slide near the stain with a marking pen. It is easy to focus on the pen mark, and you will know that you have the top of the slide in focus and can then search for the smear.

13. Record your results.

14. If you want to save your stained slide, it can be saved with the oil on it. If you do not want to save the slide, simply clean it with cleanser and water. The staining procedure kills the bacteria and the slide does not need to be boiled before cleaning.

15. **Important:** Wipe off the oil from the immersion lens with lens paper before storing the microscope.

The Negative Stain

This stain can be used to observe capsules or storage material. However, in this exercise the negative stain will be used to compare the appearance of the same organisms using the two staining procedures.

Materials

Culture
Same cultures used for simple stain
Bottle of India ink

Procedure

1. Place a drop of water on a clean slide and add organisms with a loop until the drop is cloudy.

2. Mix a loopful of India ink into the drop and spread the mixture out into a thin film.

3. Let dry and examine under the microscope. Bacteria can be seen as clear areas on a black background.

4. Record your results.

EXERCISE

Laboratory Report: Simple Stains:
Positive and Negative Stains

Results

1. Simple stain

	Staphylococcus	*Bacillus*	*Micrococcus*	*Enterococcus*

    Draw shape
    and arrangement

2. Negative stain

Questions

1. What are the advantages of a simple stain over a wet mount?

2. Do you need more or less light when viewing a stained preparation compared to a wet mount?

3. What information can you observe in a wet mount that cannot be seen in a stained preparation?

4. How does the negative stain compare to the simple stain?

5. How many μm are in a millimeter (mm)? _____
   How many μm are in a meter (m)? _____

EXERCISE  5

Multiple and Differential Stains

Getting Started

Multiple stains involve at least two dyes. They are also called differential stains because they specifically stain certain morphological features.

Gram Stain

The Gram stain is especially useful as one of the first procedures in identifying organisms because it reveals not only the morphology and the arrangement of the cells, but also information about the cell wall.

Near the turn of the century, Christian Gram devised the staining procedure when trying to stain bacteria so that they contrasted with the tissue sections he was observing. Many years later, it was found that purple (Gram-positive) bacteria had thick cell walls of **peptidoglycan,** while pink (Gram-negative) bacteria had much thinner cell walls of peptidoglycan surrounded by an additional membrane. The thick cell wall retains the purple dye in the procedure, but the thin wall does not (table 5.1).

In the Gram stain, a bacterial smear is dried and then heat-fixed to cause it to adhere to the glass slide (as in the simple stain). It is then stained with crystal violet dye, which is rinsed off and replaced with an iodine solution. The iodine acts as a **mordant**—that is, it binds the dye to the cell. The smear is then decolorized with alcohol and **counterstained** with safranin. In Gram-positive organisms, the purple crystal violet dye, complexed with the iodine solution, is not removed by the alcohol and thus the organisms remain purple. On the other hand, the purple stain is removed from Gram-negative organisms by the alcohol and the colorless cells take up the red color of the safranin counterstain.

Note: Many clinical laboratories use a 50/50 mixture of alcohol and acetone because it destains faster than 95% alcohol. If the instructor would rather not use acetone, 95% alcohol is just as effective, but the stain must be decolorized longer (up to 30 seconds).

Special Notes to Improve Your Gram Stains

1. Gram-positive organisms can lose their ability to retain the crystal violet complex when they are old. This can happen when a culture has only been incubating 18 hours—the genus *Bacillus* is especially apt to become Gram negative. Use young, overnight cultures whenever possible. It is interesting to note that Gram-positive organisms can appear Gram negative, but Gram-negative organisms almost never appear Gram positive.
2. Another way Gram-positive organisms may appear falsely Gram negative is by over decolorizing in the Gram-stain procedure. If excessive amounts of acetone/alcohol are used, almost any Gram-positive organism will lose the crystal violet stain and appear Gram negative.
3. If you are staining a very thick smear, it may be difficult for the dyes to penetrate properly. This is not a problem with broth cultures, which are naturally quite dilute, but be careful not to make the suspension from a colony in a drop of water too thick.
4. When possible, avoid making smears from inhibitory media such as eosin methylene blue (EMB) because the bacteria frequently give variable staining results and can show atypical morphology.

Table 5.1 Appearance of the Cells After Each Procedure

	Gram +	Gram −
Crystal violet	Purple	Purple
Iodine	Purple	Purple
Alcohol	Purple	Colorless
Safranin	Purple	Pink

5. The use of safranin in the Gram stain is not essential. It is simply used as a way of dying the colorless cells so they contrast with the purple. For those who are color-blind and have difficulty distinguishing pink from purple, other dyes might be tried as counterstains.

Definitions

Counterstain.   A stain used to dye unstained cells a contrasting color in a differential stain.

Mordant.   A substance that increases the adherence of a dye.

Peptidoglycan.   The macromolecule making up the cell wall of most bacteria.

Vegetative cell.   A cell that has not formed spores or other resting stages.

Objectives

1. To learn the Gram-stain procedure.
2. To learn to distinguish Gram-positive organisms from Gram-negative organisms.

References

Gerhardt, Philip, ed. *Manual for general and molecular bacteriology.* Washington, D.C.: American Society for Microbiology, 1999.

McGonagle, Lee Anne. *Procedures for diagnostic bacteriology.* 7th ed. Seattle, WA: Department of Laboratory Medicine, University of Washington, 1992.

Murray, Patrick et al. *Manual of clinical microbiology.* 7th ed. Washington, D.C.: ASM Press, 1992.

Nester et al. *Microbiology: A human perspective,* 4th ed., 2004. Chapter 3, Section 3.2.

Materials

Staining bottles of the following:
   crystal violet
   iodine
   acetone/alcohol or 95% alcohol
   safranin
Clothespin or forceps
Staining bars

Overnight cultures growing on TS agar slants
   Escherichia coli
   Bacillus subtilis
   Staphylococcus epidermidis
   Enterococcus faecalis
   Micrococcus luteus

Procedure for Gram Stain

1. Put two drops of water on a clean slide. In the first drop, make a suspension of the unknown organism to be stained just as you did for a simple stain (see preparing a smear in exercise 4). In the second drop, mix together a known Gram-positive organism and a known Gram-negative organism. This mixture is a control to ensure that your Gram-stain procedure (figure 5.1) is giving the proper results. Heat-fix the slide.
2. Place the slide on a staining bar across a sink (or can). Alternatively, hold the slide with a clothespin or forceps over a sink.
3. Flood the slide with crystal violet until the slide is completely covered. Leave it on for 6–30 seconds and then discard into the sink. The timing is not critical. Rinse the slide with water from a wash bottle or gently running tap water.
4. Flood the slide with Gram's iodine for about 12–60 seconds and wash with water.
5. Hold the slide at a 45° angle and carefully drip acetone/ethanol over it until no more purple dye runs off. Immediately wash slide with tap water. Thicker smears may take longer than thinner ones, but acetone/alcohol should usually be added for 1–2 seconds and no more than 5 seconds. Timing is critical in this step.
6. Flood the slide with safranin and leave it on for 10–30 seconds—timing is not important. Wash with tap water. Safranin is a counterstain because it stains the cells that have lost the purple dye.
7. Blot the slide carefully with a paper towel to remove the water, but do not rub from side to side. When it is completely dry, observe the slide under the microscope. Remember that you must use the oil immersion lens to

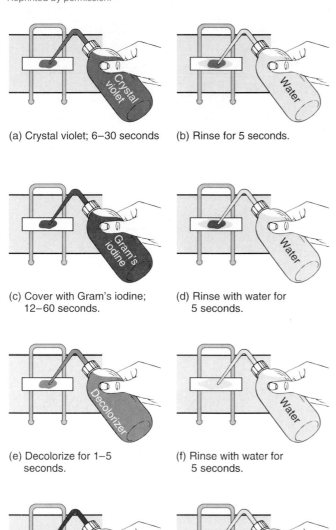

Figure 5.1 (a-i) Gram-stain procedure. From John P. Harley and Lansing M. Prescott, *Laboratory Exercises in Microbiology*, 4th ed. Copyright © 1999 The McGraw-Hill Companies. All Rights Reserved. Reprinted by permission.

(a) Crystal violet; 6–30 seconds.

(b) Rinse for 5 seconds.

(c) Cover with Gram's iodine; 12–60 seconds.

(d) Rinse with water for 5 seconds.

(e) Decolorize for 1–5 seconds.

(f) Rinse with water for 5 seconds.

(g) Counterstain with safranin; 10–30 seconds.

(h) Rinse for 5 seconds.

(i) Blot dry with a paper towel.

observe bacteria. Compare your stain to the control mixture on the same slide and with color plate 7.

8. Describe the appearance of your stained bacteria in the Results section of the Laboratory Report.

9. Be sure to remove the immersion oil from the lens with lens paper before storing the microscope.

Getting Started

Optional Stains

Acid-fast Stain

The acid-fast stain is useful for identifying bacteria with a waxy lipid cell wall. Most of these organisms are members of a group of bacteria called mycobacteria. Although there are many harmless bacteria in this group, it also includes *Mycobacterium tuberculosis*, which is the cause of tuberculosis in humans. These organisms have a Gram-positive cell wall structure, but the lipid in the cell wall prevents staining with the Gram-stain dyes.

In the Ziehl-Neelsen (Kinyoun modification) acid-fast stain procedure, the dye carbolfuchsin stains the waxy cell wall. Once the lipid-covered cell has been dyed, it cannot easily be decolorized—even with alcohol containing HCL (which is called acid-alcohol). Nonmycobacteria are also dyed with the carbolfuchsin, but are decolorized by acid-alcohol. These colorless organisms are stained with methylene blue so they contrast with the pink acid-fast bacteria that were not decolorized.

The reason this stain is important is that one of the initial ways tuberculosis can be diagnosed is by the presence of *Mycobacterium* in a patient's sputum. (Sputum is a substance that is coughed up from the lungs and contains puslike material.) Tuberculosis is a very serious disease worldwide and is now seen in the United States after decreasing for about 80 years. Since the process of finding acid-fast organisms in sputum is quite difficult and time-consuming, this test is usually performed in state health laboratories.

Objectives

1. To become familiar with acid-fast organisms.
2. To prepare an acid-fast stain.

References

Gerhardt, Philip, ed. *Manual for general and molecular bacteriology*. Washington, D.C.: American Society for Microbiology, 1994.

Nester et al. *Microbiology: A human perspective*, 4th ed., 2004. Chapter 3, Section 3.2.

Materials

Cultures

   Mycobacterium smegmatis

Carbolfuchsin in staining bottles

Methylene blue in staining bottles

Acid-alcohol in staining bottles

Beaker

Metal or glass staining bars

Procedure for Acid-Fast Stain (Kinyoun modification)

1. Prepare a smear of the material and heat-fix (see exercise 4).
2. Cover the smear with Kinyoun carbolfuchsin and stain for 3–5 minutes. Do not heat (figure 5.2).
3. Rinse with water.
4. Decolorize with acid-alcohol for 10–30 seconds.
5. Rinse with water.
6. Counterstain with methylene blue for 20–30 seconds.
7. Rinse with water.
8. Blot dry carefully and examine under the oil immersion lens.
9. Record results.

Getting Started

Differential Stains of Bacterial Cell Structures

Although bacteria have few cell structures observable by light microscopy when compared to other organisms, some have:

1. Capsules. A capsule is a somewhat gelatinous coating surrounding the cell. It can consist of amino acids or carbohydrates and it can protect the bacterium from engulfment by white blood cells. The ability to produce a

Figure 5.2 (*a-g*) Acid-fast staining procedure.

From John P. Harley and Lansing M. Prescott, *Laboratory Exercises in Microbiology*, 4th ed. Copyright © 1999 The McGraw-Hill Companies. All Rights Reserved. Reprinted by permission.

(a) Apply carbolfuchsin to smear for 5 minutes.

(b) Rinse with water.

(c) Decolorize with acid-alcohol; 10–30 seconds.

(d) Rinse with water.

(e) Counterstain with methylene blue; 20–30 seconds.

(f) Rinse with water.

(g) Blot dry with a paper towel.

capsule frequently depends on the availability of certain sugars. *Streptococcus mutans*, for example, produces a capsule when growing on sucrose, but not when growing on glucose.

2. Endospores. Some organisms such as *Bacillus* and *Clostridium* can form a resting stage called an endospore, which will protect them from heat, chemicals, and starvation. When the cell determines that conditions are becoming unfavorable due to a lack of nutrients or

moisture, it forms an endospore. Then when conditions become favorable again the spore can germinate and the cell can continue to divide. The endospore is resistant to most stains so special staining procedures are needed.

3. Storage granules. Some organisms have storage granules of phosphate, sulfur, or carbohydrate. Some of these granules can easily be seen with certain stains.

4. Bacterial flagella. Some bacteria have flagella (flagellum, singular) for motility. Their width is below the resolving power of the microscope so they cannot be seen in a light microscope (the flagella seen at each end of *Spirillum* in a wet mount is actually a tuft of flagella). Flagella can be visualized if they are dyed with a special stain that precipitates on them, making them appear much thicker. The arrangement of the flagella on bacteria is usually characteristic of the organism and can aid in identification.

Objectives

1. To become familiar with various structures and storage products of bacteria.
2. To learn various staining procedures for these structures.

References

Gerhardt, Philip, ed. *Manual for general and molecular bacteriology.* Washington, D.C.: American Society for Microbiology, 1994.

Murray, Patrick et al. *Manual of clinical micro-biology.* 7th ed. Washington, D.C.: ASM Press, 1999.

Nester et al. *Microbiology: A human perspective,* 4th ed., 2004. Chapter 3, Section 3.2.

Capsule Stain

Materials

Cultures
Klebsiella or other organism with a capsule growing on a slant
India ink

Procedure for Capsule Stain

1. Make a suspension of the organism in a drop of water on a clean slide.
2. Put a drop of India ink next to it.
3. Carefully lower a coverslip over the two drops so that they mix together. There should be a gradient in the concentration of the ink.
4. Examine under the microscope and find a field where you can see the cells surrounded by a halo in a black background.
5. Drop slides in a beaker or can of boiling water and boil for a few minutes before cleaning. This is necessary because the bacteria are not killed in the staining process.
6. Record results.

Endospore Stain

Materials

Culture
Bacillus cereus on nutrient agar slant after three or four days incubation at 30°C
Malachite green in staining bottles
Safranin in staining bottles
Metal or glass staining bars
Beaker or can

Procedure for Endospore Stain

1. Prepare a smear on a clean slide and heat-fix.
2. Add about an inch of water to a beaker and bring it to a boil.
3. Place two short staining bars over the beaker and place a slide on them.
4. Tear a piece of paper towel a little smaller than the slide and lay on top of the smear. The paper prevents the dye from running off the slide.
5. Flood the slide with malachite green and steam for 5 minutes. Continue to add stain to prevent the dye from drying on the slide (figure 5.3).
6. Decolorize with water for about 30 seconds by flooding with water or holding under gently

Figure 5.3 (a-e) Procedure for staining endospores. From John P. Harley and Lansing M. Prescott, *Laboratory Exercises in Microbiology*, 4th ed. Copyright © 1999 The McGraw-Hill Companies. All Rights Reserved. Reprinted by permission.

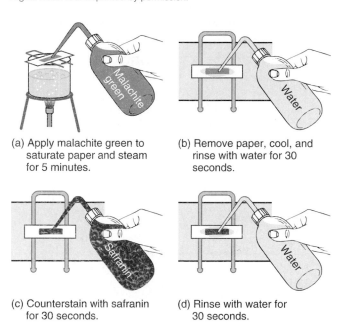

(a) Apply malachite green to saturate paper and steam for 5 minutes.

(b) Remove paper, cool, and rinse with water for 30 seconds.

(c) Counterstain with safranin for 30 seconds.

(d) Rinse with water for 30 seconds.

(e) Blot dry with a paper towel.

running tap water. The **vegetative cells** (dividing cells) will lose the dye, but the endospores will retain the dye.

7. Counterstain with safranin for about 30 seconds and then wash with tap water for 30 seconds. Blot dry carefully.

8. Observe with the oil immersion lens. The endospores will appear green and the vegetative cells will appear pink. Sometimes the endospore will still be seen within the cell, and its shape and appearance can be helpful in identifying the organism. In other cultures, the endospores may be free because the cells around them have disintegrated (figure 5.4).

9. Record results.

10. Prepare and observe a Gram stain of the same culture (optional).

Figure 5.4 Appearance of endospores stained with spore stain and Gram stain. Note: The *Bacillus* frequently lose their ability to stain Gram positive.

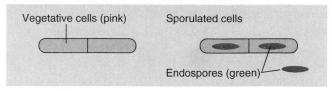

Spore Stain of *Bacillus* with Malachite Green

Vegetative cells (pink)   Sporulated cells

Endospores (green)

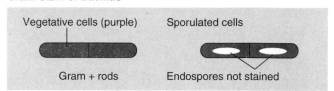

Gram Stain of *Bacillus*

Vegetative cells (purple)   Sporulated cells

Gram + rods   Endospores not stained

Note: When bacteria containing endospores are Gram stained the endospores do not stain and the cells appear to have holes in them. (See figure 5.4.)

Storage Granules Stain

Many organisms can store materials that are abundant in their environment for use in the future. For example, phosphate can be stored as metachromatic granules (also called volutin granules). When organisms containing these granules are stained with methylene blue, the phosphate granules are stained a darker reddish blue.

Materials

Cultures
    Spirillum grown in nutrient broth
Methylene blue in staining bottles

Procedure for Storage Granules Stain

1. Prepare a smear from the broth. It might be helpful to remove the organisms from the bottom of the tube with a capillary pipet. Place a drop on a clean slide. Dry and heat-fix.

2. Flood the slide with methylene blue for about 20–30 seconds.

3. Wash with tap water and blot dry.

4. Observe with the oil immersion lens. The metachromatic granules should appear as dark reddish-blue bodies within the cells.
5. Record results.

Flagellar Stain

There are three basic kinds of arrangement of flagella:

Definitions

Lophotrichous.   A tuft of flagella at one or each end of the organism, as in *Spirillum*.

Peritrichous.   The surface of the organism is covered with flagella, such as *E. coli*.

Polar.   A single flagellum at one or both ends of the organism such as *Pseudomonas*.

Materials

Stained demonstration slides of *Escherichia coli* and *Pseudomonas*

Procedure for Flagellar Stain

1. Observe flagellar stained slides of several organisms and note the pattern of flagella. It is difficult to perform this staining procedure, so prestained slides are recommended.
2. Record results.

NOTES:

Name _____ Date _____ Section _____

EXERCISE

 **5**

Laboratory Report: Multiple and Differential Stains

Results

	Gram Reaction	Arrangement (sketch)
E. coli		
B. subtilis		
S. epidermidis		
E. faecalis		
M. luteus		

Optional Stains	Organism	Appearance
Acid-fast		
Capsule		
Endospore		
Storage granules		
Flagella		

Questions

1. What is the function of each one of the Gram-stain reagents?

2. Give two reasons Gram-positive organisms sometimes appear Gram-negative.

3. What is the purpose of using a control in the Gram stain?

4. What is a capsule?

5. What are storage granules and why are they important to the cell?

6. How does an endospore appear (draw and indicate color):

    a. when Gram stained?

    b. when spore stained?

7. What is another way you could determine whether an organism was motile besides observing a flagellar stain?

8. Why can't you Gram stain an acid-fast organism?

NOTES:

*I*NTRODUCTION to Microbial Growth

In order to study microorganisms, you must be able to grow or culture them. One bacterium is too small to do anything that can be easily measured, but a whole population of bacteria produce an effect big enough to be readily seen or counted. It is important that the population (culture) contain just one kind of organism. Such a culture is called a pure culture and is defined as a population of bacteria that have all grown from a single cell.

It might appear to be a very difficult problem to separate out one single bacterium from the millions of others and then permit it alone to form its own colony. Fortunately, there is a simple technique called the streak plate method, which spreads individual bacteria on an agar plate. Colonies that grow from the widely separated bacteria are far enough apart that they can be easily transferred and studied further.

Organisms in the laboratory are frequently grown either in a broth culture or on a solid agar medium. A broth culture is useful for growing large numbers of organisms. Agar medium is used in a petri dish when a large surface area is important, as in a streak plate. On the other hand, agar medium in tubes (called slants) is useful for storage because the small surface area is not as easily contaminated and the tubes do not dry out as fast as plates. You will be able to practice using media in all these forms (figure I.3.1).

Another important skill is the ability to prevent other bacteria from growing in the pure culture you are studying. Aseptic technique is a set of procedures designed to: (1) prevent a culture from being contaminated and (2) prevent the culture from contaminating you or your surroundings.

You will also use different kinds of media in this section. Most media are formulated so that they will support the maximum growth of various organisms, but some media have been designed to permit the growth of desired organisms and inhibit others (selective). Still other media have been formulated to change color or in some other way distinguish one colony from another (differential). These media can be very useful when trying to identify an organism.

Figure I.3.1 (*a-e*) Diagram of different media in different forms. From John P. Harley and Lansing M. Prescott, *Laboratory Exercises in Microbiology*, 5th ed. Copyright © 2002 The McGraw-Hill Companies. All Rights Reserved. Reprinted by permission.

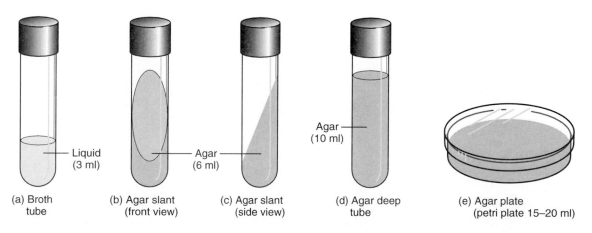

(a) Broth tube — Liquid (3 ml)

(b) Agar slant (front view) — Agar (6 ml)

(c) Agar slant (side view)

(d) Agar deep tube — Agar (10 ml)

(e) Agar plate (petri plate 15–20 ml)

It is also important to know how to count bacteria. You will have an opportunity to learn about several techniques and their advantages and disadvantages.

In the next set of exercises, you will learn how to isolate a pure culture, to use aseptic technique, and to grow and count microorganisms. You will also be introduced to various kinds of media that are formulated for different purposes.

In these exercises, no pathogenic organisms are used, but it is very important to treat these cultures as if they were harmful because you will be able to then work safely with actual pathogens. Also, almost any organism may cause disease if there are large numbers in the wrong place.

EXERCISE
6 Pure Culture and Aseptic Technique

Getting Started

Aseptic Technique

The two goals of **aseptic** (or **sterile**) technique are to prevent contamination of your culture with organisms from the environment and to prevent the culture from contaminating you or others.

In this exercise, you will transfer sterile broth back and forth from one tube to another using aseptic technique. The goal is to do it in such a way that you will not permit any organism in the environment from entering the tubes. You will be using both a sterile pipet and a flamed loop on the same set of broths. After you have practiced aseptically transferring the broth, you will **incubate** the broth tubes for a few days to determine if they are still sterile. If you used good technique, the broth will still be clear; if organisms were able to enter from the environment, the broth will be cloudy from the bacterial growth.

When you can successfully transfer sterile broth aseptically without contamination, you can use the same technique to transfer a pure culture without contaminating it or the environment.

Definitions

Aseptic.   Free of contamination.

Incubate.   Store cultures under conditions suitable for growth, often in an incubator.

Sterile.   Aseptic; free of viable bacteria or viruses.

Objectives

1. To learn aseptic technique procedures and their importance.
2. To learn to isolate colonies using a streak plate technique.

Reference

Nester et al. *Microbiology: A human perspective,* 4th ed., 2004. Chapter 4, Section 4.1.

Broth-to-Broth Transfer with a Wire Loop

Materials

Per student
  Tubes of TS broth, 2
  Inoculating loop

Procedure

1. Always label tubes before adding anything to them. In this exercise, you will be transferring sterile broth from one tube to another, so that both tubes will have the same label; however, in general, labeling tubes before inoculation prevents mistakes.
2. Grip the loop as you would a pencil and flame the wire portion red hot. Hold it at an angle so that you will not burn your hand (figure 6.1).
3. After the loop has cooled for a few seconds, pick up a tube in the other hand and remove the cap of the tube with the little finger (or the fourth and little fingers) of the hand holding the loop.
4. Flame the mouth of the tube by passing it through a Bunsen burner flame and then use the sterile loop to obtain a loopful of liquid from the tube. Flame the mouth of the tube and replace the cap. If you have trouble picking up a loopful of material, check to be sure that your loop is a complete circle without a gap.
5. Set down the first tube and pick up the second tube. Remove the cap, flame it, and

Figure 6.1 Aseptic technique for removing a loopful of broth culture. (a) Hold the culture tube in your left hand and the loop in your right hand (reverse if you are left-handed). Flame the loop to sterilize it. (b) Remove the culture tube plug or cap, and flame the mouth of the culture tube. (c) Insert the sterile loop into the culture tube. (d) Remove the loopful of inoculum, and flame the mouth of the culture tube again. (e) Replace the culture tube plug or cap. Place the culture tube in a test tube rack. (f) Reflame the loop.

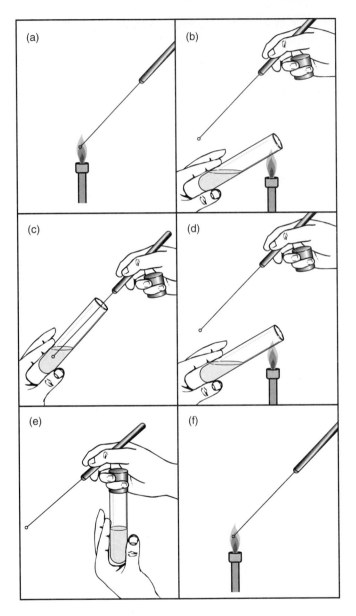

deposit a loopful of material into the liquid of the second tube. Withdraw the loop, flame the tube, and then replace the cap. Be sure to flame the loop before setting it on the bench (your loop would normally be contaminated

with the bacteria you were inoculating). It is usually convenient to rest the hot loop on the edge of the Bunsen burner.

6. When learning aseptic technique, it is better to hold one tube at a time; later, you will be able to hold two or three tubes at the same time.

Transferring Broth with a Pipet

Note: Sterile pipets are used when it is necessary to transfer known amounts of material. Some laboratories use plastic disposable pipets and others use reusable glass pipets. Be sure to follow the instructor's directions for proper disposal after use (never put a used pipet on your bench top). Mouth pipetting is dangerous and is not permitted. A variety of bulbs or devices are used to draw the liquid up into the pipet and your laboratory instructor will demonstrate their use (figure 6.2).

The same broth tubes used for practice with the loop can be used to practice pipetting the broth back and forth.

Materials

TS broth tubes from previous procedure
1-ml pipet
Bulb or other device to fit on end of the pipet

Procedure

First Session

1. Open a sterile pipet at the top, insert a bulb on the end, then carefully remove the pipet from the package or canister without touching the tip. Grip the pipet as you would a pencil. The pipet is plugged with cotton to filter the air going into it. Discard the pipet if liquid inadvertently wets the plug—air will no longer enter the pipet and the measured liquid will not flow out. Notify your instructor if the bulb is contaminated.

2. Pick up a tube with your other hand and remove the cap with the little finger of the hand holding the pipet. Flame the tube. Expel air from the rubber bulb and insert the pipet

Figure 6.2 Various devices for filling pipets.

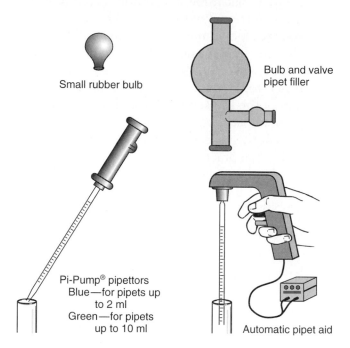

Small rubber bulb

Bulb and valve pipet filler

Pi-Pump® pipettors
Blue—for pipets up to 2 ml
Green—for pipets up to 10 ml

Automatic pipet aid

tip into the liquid. Note that the liquid must be drawn to the 0 mark for 1 ml when using a 1-ml pipet. Draw the liquid up to the desired amount, remove the pipet, flame the tube, replace the cap, and then put the tube back in the rack.

3. Pick up the second tube and repeat the steps used with the first tube except that the liquid is expelled into the tube.
4. Repeat the above steps with the same tubes until you feel comfortable with the procedure.
5. Dispose of the pipet as directed.
6. Incubate the tubes until the next period at 37°C.

Second Session

1. Observe the tubes of broth for turbidity. If they are cloudy, organisms contaminated the broth during your practice and grew during incubation. With a little more practice, you will have better technique. If the broths are clear, there was probably no contamination and you transferred the broth without permitting the entry of any organisms into the tubes.
2. Record results.

Streak Plate Technique

Materials

Per student
> Trypticase soy agar (TSA) plates, 2

Cultures
> Broth culture containing a mixture of two organisms such as *Micrococcus* and *Staphylococcus*

Procedure

First Session

1. Label the agar plate on the bottom with your name and date.
2. Divide the plate into three sections with a "T" as diagrammed (figure 6.3).
3. Sterilize the loop in the flame by heating the whole length of the wire red hot. Hold it at an angle so you do not heat the handle or roast your hand.
4. Gently shake the culture to be sure both organisms are suspended. Aseptically remove a loopful of the culture and holding the loop as you would a pencil, spread the bacteria on section 1 of the plate by streaking back and forth. The more streaks, the better chance of an isolated colony. As you work, partially cover the petri dish with the cover to minimize organisms falling on the plate from the air. Use a gliding motion and avoid digging into the agar. Don't press the loop into the surface. If your loop is not smooth or does not form a complete circle, it can gouge the agar and colonies will run together. Note that you can see the streak marks if you look carefully at the surface of the plate.
5. Burn off all the bacteria from your loop by heating it red hot. This is very important because it eliminates the bacteria on your loop. Wait a few seconds to be sure the loop is cool.
6. Without going into the broth again, streak section 2 (see figure 6.3) of the petri plate. Go into section 1 with about three streaks and spread by filling section 2 with closely spaced streaks.

Figure 6.3 (*a-g*) Preparation of a streak plate.

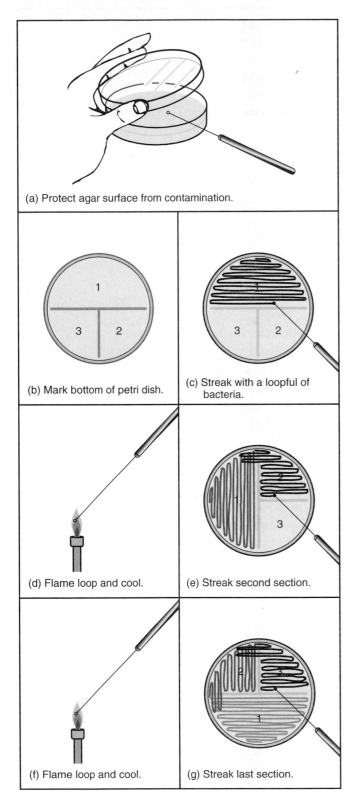

(a) Protect agar surface from contamination.

(b) Mark bottom of petri dish.

(c) Streak with a loopful of bacteria.

(d) Flame loop and cool.

(e) Streak second section.

(f) Flame loop and cool.

(g) Streak last section.

7. Again heat the loop red hot. Go into section 2 with about three streaks and spread by filling section 3 with streaks. The more streaks you are able to make, the greater will be your chance of obtaining isolated colonies.

8. Heat loop red hot before placing it on the bench top. Usually you can rest it on some part of the Bunsen burner so that it can cool without burning anything.

9. Repeat the procedure with a second plate for additional practice.

10. Incubate the plates in the 37°C incubator.

Second Session

Observe your streak plates and record results.

EXERCISE

Laboratory Report: Pure Culture and Aseptic Technique

Results

	Clear	Turbid
Tube 1		
Tube 2		

	Number of Colonies Isolated
Streak plate 1	
Streak plate 2	

Did you obtain isolated colonies of each culture?

Questions

1. What is the definition of a pure culture?

2. Why is sterile technique important? Give two reasons.

3. What is the purpose of a streak plate?

4. Why is it important to avoid digging into the agar with the loop?

5. Is there anything you can do to improve your streak plate technique?

EXERCISE

7

Defined, Undefined, Selective, and Differential Media

Getting Started

Microbiologists have developed several different types of culture media for a variety of different purposes. If a medium is totally made up of specific amounts of chemicals it is called a **defined medium.** If, however, it contains such mixtures as yeast extract or enzymatic digests of protein, it is termed an **undefined medium,** or complex medium, because the exact amount and kinds of large organic molecules are not known.

Undefined media are frequently called **rich media.** These media tend to support the growth of bacteria better because they contain more preformed nutrients, and the organisms do not have to use energy and materials to synthesize the compounds supplied in the medium. Many bacteria can grow only on this kind of medium because they cannot synthesize all the necessary components needed for growth and must be provided with preformed amino acids, vitamins, and other growth factors. Some organisms require only one or two vitamins or amino acids while other bacteria need many complex growth factors, and are termed fastidious.

Two other valuable types of media are **selective** and **differential media.** Frequently, it is important to isolate one organism in a mixture of bacteria. Normal flora can contaminate the culture; for example, a wound culture may be mixed with large amounts of *Staphylococcus* normally found on the skin. It can be difficult to isolate or even detect the pathogenic organism among all the nonpathogenic organisms present. Selective media have been designed to permit some bacteria to grow but not others, so that certain bacteria can be isolated even if they constitute only a small percentage of the population.

Differential media are also useful for isolating and identifying bacteria. By observing the appearance of colonies growing on this agar, it is possible to determine characteristics such as whether or not they can ferment certain sugars.

In this exercise, you will have an opportunity to observe the growth of organisms on three different media.

Glucose Salts Agar　This is a simple defined medium. Only organisms that can make all their cellular components from glucose and inorganic salts are able to grow on it.

Trypticase Soy Agar　This is a rich, undefined medium made from an enzymatic digest of protein and soy product. Organisms that require vitamins or other growth factors are able to grow on it.

EMB (Eosin Methylene Blue) Agar　This is a selective medium permitting the growth of Gram-negative enteric rods and inhibiting the growth of Gram-positive bacteria. In addition, the medium is also differential because it contains the sugar lactose. Organisms that can ferment lactose produce purple colonies and those that cannot, produce white or very light pink colonies. The colonies of *E. coli*, a lactose fermenter, are dark purple. They also give the medium a distinctive metallic green sheen caused by the large amounts of acid produced. The colonies of *Enterobacter*, also a lactose fermenter, usually are more mucoid with purple centers. (Mucoid colonies have a slimy appearance.)

Definitions

Defined medium.　A synthetic medium composed of inorganic salts and usually a carbon source such as glucose.

Differential medium.　Medium permitting certain organisms to be distinguished from others by the appearance of their colonies.

Rich (or **enriched**) **medium.** A medium containing many growth factors. Usually it is an undefined medium made from meat, plant, or yeast extracts.

Selective medium. Medium formulated to permit the growth of certain bacteria but not others.

Undefined medium. A complex medium in which the exact amounts of components and their composition are unknown because it is made of extracts or enzymatic digests of meat, plants, or yeast.

Objectives

1. To compare the growth of organisms on a defined and a rich medium.
2. To compare the growth of organisms on a selective and a differential medium.
3. To understand the relationship between the growth of an organism and the composition of the medium.

Reference

Nester et al. *Microbiology: A human perspective,* 4th ed., 2004. Chapter 4, Section 4.5.

Materials

Cultures growing in trypticase soy broth

    Escherichia coli

    Staphylococcus epidermidis

    Pseudomonas aeruginosa

    Enterobacter aerogenes

Media per team

Trypticase soy (TS) agar plate

Glucose mineral salts agar plate

EMB (eosin methylene blue) agar plate

Procedure

First Session

1. With a marking pen, divide the bottom of the petri plates into quadrants. Label the plates with your name and date. Label each quadrant with the organism as shown in figure 7.1.
2. Inoculate each quadrant of the plate with a loopful of the culture in a wavy line.
3. Invert and incubate at 37°C for 48 hours.

Second Session

1. Observe and compare the growth on the three plates.
2. Record the results.

Figure 7.1 A diagram of the labeled media plates.

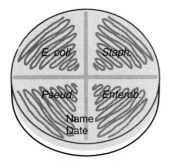

Trypticase soy

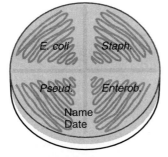

Glucose mineral

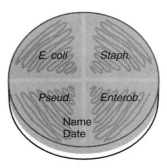

Eosin methylene blue (EMB)

EXERCISE

Laboratory Report: Defined, Undefined, Selective, and Differential Media

Results

	Glucose Salts	Trypticase Soy (TS)	Eosin Methylene Blue (EMB)
E. coli			lac–/lac+
Staphylococcus			
Pseudomonas			
Enterobacter			

Indicate the amount of growth:
   0=no growth
   +=slight growth
  ++=good growth
 Lac –=no lactose fermentation on EMB
 Lac +=lactose fermentation on EMB

Questions

1. Which organisms could not grow on the glucose salts medium? Which organisms could grow on it?

2. Which organisms do not require any growth factors?

3. Of the organisms that could grow on both TS agar and glucose salts, did some organisms grow better on the TS agar than the glucose salts? Can you propose a reason?

4. Which organisms could grow on the EMB agar?

5. Which organisms could ferment lactose?

6. Could you differentiate *E. coli* from other organisms growing on EMB? How?

7. In general, EMB selects for what kind of organisms?

8. What kinds of organisms does EMB differentiate?

EXERCISE

8 Quantification of Microorganisms

Getting Started

It is frequently important to count bacteria. For example, you may want to know the number of bacteria in a sample of raw chicken or the number of bacteria per ml of water in a swimming pool. Special techniques have been devised to enumerate bacteria, each with advantages and disadvantages. Three common methods are discussed.

Plate Count This method is based on the premise that each **viable** bacterium will produce a colony when growing on an agar plate. A sample of the material to be counted is suspended in liquid and placed in an empty petri plate. Melted, cooled agar is then added to the plate and mixed with the inoculum. After incubation, each organism produces a colony in the agar that can then be counted. The plate count is used very frequently but it has advantages and disadvantages that should be considered prior to use. Some of these are discussed next.

1. Bacteria are usually present in very large numbers—an overnight broth culture of *E. coli* can easily contain one billion cells/ml. However, the maximum number of colonies that can be accurately counted on a plate is usually set at 300. Therefore, most samples must be diluted to low enough numbers that the plates will have distinct colonies that can be counted. Since it is usually not possible to know exactly how dilute to make a sample to obtain a countable plate, several different dilutions must be plated.
2. Some bacteria tend to stick together; therefore, sometimes two or more bacteria will give rise to one colony. This gives results of a lower number than are actually present.
3. If a sample has many different kinds of bacteria, it is not possible to have a medium or conditions that will support all their various necessary growth conditions. Soil, for

example, may contain organisms that will not grow unless the temperature is above 50°C; in contrast, other organisms are inhibited at these temperatures. These problems must be considered when a sample of mixed bacteria is enumerated.

4. It can take at least 24 hours to obtain the results of a plate count.
5. The plate count method does have two advantages over other methods, however. Only viable organisms are counted, which are the ones usually considered important. Also, samples with small numbers can be counted, which would have insufficient numbers for other methods.

Direct Count In this method of counting organisms, a suspension of bacteria is placed on a slide that has been ruled into squares and is designed to hold a specific volume of liquid. By counting the bacteria that appear on the grid areas, the number of organisms in the sample can be calculated. It is a much faster test than the plate count, but it does have several drawbacks. First, there must be about 1×10^7 organisms/ml before there are enough to be seen, and second, viable and nonviable organisms appear the same under a microscope.

Turbidometric In this method, a spectrophotometer is used to measure the **turbidity** or **optical density (O.D.)** of bacteria in a broth. The more bacteria, the cloudier the broth and the higher the optical density. In this method, you must first correlate plate counts with optical density readings. This must be done with each strain of bacteria because organisms are different sizes. For instance, an optical density reading of 0.2 for a broth culture of one *E. coli* strain is equal to 1×10^8 cells/ml. The same number of another organism would have a different optical density. Once the correlation between O.D. and plate counts has been determined, the correlation can be used as an extremely convenient

method of determining numbers of organisms. This method is used to determine the generation time in exercise 10. Other methods of enumerating bacteria are discussed in exercise 33, which is on water analysis.

The plate count method is used in this exercise to count the number of organisms in two broth cultures: one turbid (sample A) and the other with no visible turbidity (sample B). There are two methods of preparing plate counts: pour plates and spread plates. In the pour plate method (as just described), a sample is mixed with melted agar in a petri plate and colonies appearing in and on the agar are counted. In the spread plate method, a small sample is placed on the surface of the agar plate and spread with a bent glass rod so that all the colonies appear on the surface of the plate. The spread plate method is used in exercise 17, Bacterial Conjugation.

Definitions

Optical density (O.D.). A measure of the amount of turbidity. Frequently also called absorbance.

Serial dilution. Preparing a dilution in steps instead of one dilution.

Turbid. Cloudy.

Viable (bacteria). Capable of growing and dividing.

Objectives

1. To enumerate bacteria using the plate count method.
2. To understand the use of dilutions.
3. To gain insight into the number of organisms that can be present in a clear liquid and a turbid liquid.

Reference

Nester et al. *Microbiology: A human perspective,* 4th ed., 2004. Chapter 4, Section 4.2.

Note: When serially transferring concentrated suspensions to less concentrated suspensions, a new pipet must be used for each transfer to prevent carryover of organisms. (In this exercise, the instructor

may choose to use each pipet twice to conserve materials.) However, when going from less concentrated to more concentrated suspensions, the same pipet may be used with no significant error because only a few organisms will be added to the much larger sample.

Materials

For teams using sample A (turbid suspension)
  99-ml water blank, 1
  9.9-ml water blanks, 2
  9.0-ml water blanks, 3
  1-ml pipets, 7
  TS agar deeps, 4
  Sterile petri dishes, 4
  Suspension A: An overnight TS broth culture (without shaking) of *E. coli* diluted 1:1 with TS broth

Procedure for Team A

First Session

1. Label all water blanks with the dilution, as shown in figure 8.1.
2. Melt 4 TS agar deeps and hold at 50°C. It is very important not to let the deeps cool much lower than 50°C because the agar will harden and will have to be heated to boiling (100°C) before it will melt again (figure 8.2).
3. Make **serial dilutions** of the bacterial suspension.
   a. Mix the bacterial suspension by rotating between the hands and transfer 1.0 ml of the suspension to the 99-ml water blank labeled 10^{-2}. Discard the pipet.
   b. Mix well and transfer 0.1 ml of the 10^{-2} dilution to the 9.9-ml water blank labeled 10^{-4}. Discard pipet.
   c. Mix well and transfer 0.1 ml of the 10^{-4} dilution to the 9.9-ml tube labeled 10^{-6}. Discard pipet.
   d. Mix and transfer 1.0 ml of the 10^{-6} dilution to the 9.0-ml tube labeled 10^{-7}. (Note change from 0.1 ml to 1.0 ml.) Discard pipet.

Figure 8.1 Dilution scheme for Team A.

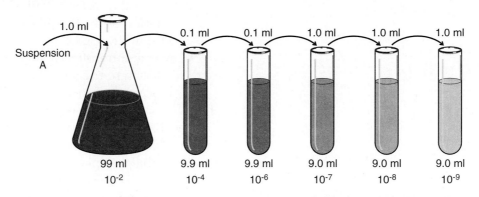

Team A

| | 1.0 ml | 0.1 ml | 0.1 ml | 1.0 ml | 1.0 ml | 1.0 ml |

Suspension A

| 99 ml | 9.9 ml | 9.9 ml | 9.0 ml | 9.0 ml | 9.0 ml |
| 10^{-2} | 10^{-4} | 10^{-6} | 10^{-7} | 10^{-8} | 10^{-9} |

Figure 8.2 (*a-f*) Melting and pouring agar deeps.

(a) Boiling water bath

Thermometer

(b) 50°C water bath

Test tube rack

(c) Wipe tube with paper towel.

(d) Flame the top of the tube after removing cap.

(e) Pour the agar into the petri dish bottom.

Agar

(f) After cooling, store in an inverted position.

e. Mix and transfer 1.0 ml of the 10^{-7} dilution to the 9.0-ml tube labeled 10^{-8}. Discard pipet.

f. Mix and transfer 1.0 ml of the 10^{-8} dilution to the 9.0-ml tube labeled 10^{-9}. Discard pipet.

4. Place samples of the dilutions 10^{-9}, 10^{-8}, 10^{-7}, and 10^{-6} into sterile labeled petri plates as follows:

a. Mix the 10^{-9} dilution and transfer 1.0 ml into a sterile petri plate labeled 10^{-9}. Add a tube of melted agar (wipe off the outside of the tube before pouring) and swirl gently by moving the plate in a figure eight pattern on the bench. Do not discard pipet. (See figure 8.2.)

b. Mix the 10^{-8} dilution and with the same pipet transfer 1.0 ml to the petri plate labeled 10^{-8}. Add melted agar and mix.

c. Mix the 10^{-7} dilution again and using the same pipet transfer 1.0 ml to the petri plate labeled 10^{-7}. Add melted agar and mix.

d. Mix the 10^{-6} dilution and using the same pipet transfer 1.0 ml to the petri plate labeled 10^{-6}. Add melted agar and mix. Discard the pipet.

5. Invert the plates after you are sure the agar has hardened (about 5 minutes), and incubate at 37°C.

Materials

For teams using sample B (a nonturbid suspension)

99-ml water blank, 1

9.9-ml water blank, 1

9.0-ml water blanks, 4

1-ml pipets, 5

TS agar deeps, 4

Sterile petri dishes, 4

Suspension B: a nonturbid suspension of bacteria

Procedure for Team B

First Session

1. Label all water blanks and plates with the dilution (figure 8.3).
2. Melt 4 TS agar deeps and hold at 50°C. It is very important not to let the agar deeps cool much lower than 50°C because the agar will harden and will have to be heated to boiling (100°C) before it will melt again.
3. Make serial dilutions of the bacterial suspension.
   a. Mix the bacterial suspension and transfer 1.0 ml of the suspension to the 99-ml water blank labeled 10^{-2}. Discard the pipet.
   b. Mix well and transfer 0.1 ml of the 10^{-2} dilution to the 9.9-ml water blank labeled 10^{-4}. (Note that this is the only time you will use a 0.1 ml-sample.) Discard pipet.
   c. Mix well and transfer 1.0 ml of the 10^{-4} dilution to the 9.0-ml tube labeled 10^{-5}. Discard the pipet.
   d. Mix and transfer 1.0 ml of the 10^{-5} dilution to the 9.0-ml tube labeled 10^{-6}. Discard the pipet.
   e. Mix and transfer 1.0 ml of the 10^{-6} dilution to the 9.0-ml tube labeled 10^{-7}. Discard the pipet.
4. Place samples of the dilutions 10^{-7}, 10^{-6}, 10^{-5}, and 10^{-4} into the pour plates as follows.
   a. Mix the 10^{-7} dilution tube and remove 1 ml to the petri plate labeled 10^{-7}. Pour the melted, cooled agar into the plate (wipe off any water on the outside of the tube before pouring the agar). (See figure 8.2.) Mix by gently moving the plate in a figure eight on the bench. Do not discard the pipet.
   b. Repeat the procedure for the 10^{-6}, 10^{-5}, and 10^{-4} dilutions using the same pipet.
5. After the agar has hardened (about 5 minutes) invert the plates and incubate at 37°C.

Second Session for Both A and B Teams

1. Count the colonies in the agar plates. Use a marking pen on the bottom of the plate to dot the colonies as you count them. Colonies growing in the agar tend to be lens-shaped and smaller than those growing on the surface but all are counted equally. If there are more than 300 colonies on the plate, label it TNTC—too numerous to count.

Figure 8.3 Dilution scheme for Team B.

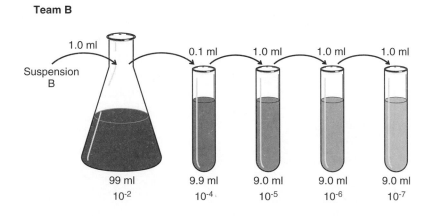

Team B

2. Choose the plate that has between 30 and 300 colonies (less than 30 gives results with a high sample error). Calculate the number of organisms/ml using the following formula:

number of organisms on the plate × 1/sample volume × 1/dilution = the number of organisms/ml in the original suspension.

Note: In this exercise the volume of all the samples is 1.0 ml.

3. Record your results. Post your results on the blackboard so that average numbers of organisms/ml for Suspension A and Suspension B can be calculated.

Understanding Dilutions

(See also Appendix 2.)

1. To make a dilution use the following formula:

sample/(diluent + sample) = the dilution

Example 1 How much is a sample diluted if 1 ml is added to 9.0 ml of water (the water is sometimes called a diluent)?

$1/(1 + 9) = 1/10$ (also expressed 10^{-1})

Example 2 How much is a sample diluted if 0.1 ml is added to 9.9 ml of water?

$0.1/(0.1 + 9.9) = 0.1/10$
$= 1:100$ or 10^{-2}

2. When a sample is serially diluted, multiply each dilution together for the final dilution. The final dilution in tube B is 1:100 or 10^{-2}.

3. To calculate the number of organisms in the original suspension use the formula:

The number of organisms/ml in the original sample = number of colonies on plate × 1/volume of sample × 1/dilution

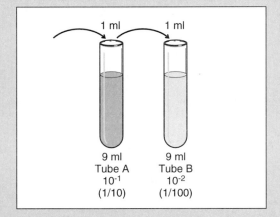

Example 3 Suppose you counted 120 organisms on a plate diluted 10^{-2}. The sample size was 0.1 ml.

Solution: 120 (number of organisms on plate) × 1/0.1 × $1/10^{-2} = 120 \times 10 \times 100$
$= 120 \times 10^3$ or 1.2×10^5 organisms/ml

Example 4 Suppose you counted 73 colonies on the plate marked 10^{-6}. If the sample size is 1.0 ml then

$73 \times 1/1.0 \times 1/10^{-6} = 73 \times 1 \times 10^6 = 73 \times 10^6$ or 7.3×10^7
organisms/ml in the original suspension

(It is important to label the answer "per ml.")

NOTES:

EXERCISE

Laboratory Report: Quantification of Microorganisms

Results

1. Which plate (dilution) had between 30 and 300 colonies?

2. How many colonies did you count?

3. How many organisms/ml were in the original suspension? (From questions 1 and 2)

	Suspension A	Suspension B
Appearance of broth		
Number of organisms/ml (class average)		
Number of organisms/ml (your data)		

Questions

1. From these results, about how many organisms/ml can be in a cloudy broth? (Show calculations.)

2. From these results, about how many organisms/ml can be in a clear broth without showing any sign of turbidity? (Show calculations.)

3. What are two sources of error in this procedure?

4. If you serially dilute a sample with three 1:10 dilutions, what is the final dilution of the last tube?

5. If you add 1.0 ml to 99 ml of water, what is the dilution of the sample?

6. If you had a solution containing 6,000 organisms/ml, how could you dilute and plate a sample so that you had a countable plate?

INTRODUCTION to the Environment and Microbial Growth

An organism cannot grow and divide unless it is in a favorable environment. Environmental factors include temperature, availability of nutrients, moisture, oxygen, salinity, osmotic pressure, and presence of toxic products.

Each bacterial species has its own particular set of optimal conditions that allows maximum growth. These conditions probably reflect the environment in which the organism grows and competes with other organisms. In the next two exercises, we examine the effects of temperature and atmosphere (oxygen) on the growth of bacteria. (Osmotic pressure is examined in exercise 13.)

NOTES:

EXERCISE

9

Aerobic and Anaerobic Growth

Getting Started

All the animals we are familiar with, including humans, have an absolute requirement for oxygen. It seems rather surprising then that there are groups of organisms that cannot grow or are even killed in the presence of oxygen. Still other kinds of organisms can grow either with or without oxygen. These three groups are classified as follows:

1. *Obligate aerobes*   Organisms that have an absolute requirement for oxygen. *Micrococcus*, a member of the skin flora, and *Pseudomonas*, a soil organism (and occasional pathogen), are obligate aerobes.
2. *Obligate anaerobes*   Organisms that cannot grow in the presence of oxygen. There are, however, varying degrees of sensitivity to oxygen. Some bacteria such as the methanogens that produce methane gas in swamps are killed by a few molecules of O_2 while others, such as *Clostridium*, usually can survive in O_2 but cannot grow until conditions become anaerobic.
3. *Facultative anaerobes*   Organisms that can grow either in the presence or absence of oxygen. (Usually they are simply called facultative.) *Escherichia coli*, a member of the intestinal flora, is facultative.

There are also other categories such as **microaerophilic** organisms, which do best in reduced amounts of oxygen, and organisms that prefer more CO_2 than the amount normally found in the atmosphere. In this exercise, however, we examine the oxygen requirements of an obligate aerobe, an obligate anaerobe, and a facultative anaerobe. Try to identify them by growing each culture **aerobically** on a slant and **anaerobically** in an **agar deep.**

Definitions

Aerobic.   In the presence of air. Air contains about 20% oxygen.

Agar deep.   A test tube filled with agar almost to the top.

Anaerobic.   In the absence of air.

Microaerophilic.   Reduced amounts of air.

Objectives

1. To understand how microbes differ in their oxygen requirements.
2. To learn to distinguish between aerobes, anaerobes, and facultative anaerobes.

Reference

Nester et al. *Microbiology: A human perspective,* 4th ed., 2004. Chapter 4, Section 4.3.

Materials

Per team
  TS broth cultures labeled A, B, and C
    Escherichia coli
    Micrococcus
    Clostridium
  TS + 0.5% glucose agar deeps, 3
  TS agar slants, 3

Procedure

First Session

1. Put the agar deeps in a beaker, or can, and fill it with water to the height of the agar. Boil

Figure 9.1 Boiling agar deeps for 10 minutes to drive out dissolved oxygen.

the tubes for 10 minutes. This will not only melt the agar but also drive out all the dissolved oxygen (figure 9.1).

After the agar hardens, air will gradually diffuse into the tube so that about the top several millimeters of the agar will be aerobic, but the remainder of the tube will be anaerobic.

2. Cool the agar in a 50°C water bath for about 10 minutes (check the temperature of the water with a thermometer). Be careful that the agar does not cool much lower than 50°C

or it will solidify and you will have to boil it to melt it again. The tube will feel hot, but you will be able to hold it.

3. Label the tubes and slants.
4. Inoculate a melted agar deep with a loopful of culture A (figure 9.2) and mix by rolling between the hands. Permit the agar to harden. This technique is often called a shake tube.
5. Inoculate a slant with a loopful of culture A by placing a loopful of broth on the bottom of the slant and making a wiggly line on the surface to the top of the slant.
6. Repeat with cultures B and C.
7. Incubate all slants and deeps at least 48 hours at 25°C. Some cultures grow so vigorously at 37°C that the gas produced blows apart the agar.

Second Session

1. Observe the surface of the slants and of the deeps, and record the growth. Compare to figure 9.3.
   Note: Sometimes the anaerobes seep down and grow between the agar slant and the walls of the glass tube where conditions are anaerobic, but not on the surface of the slant, which is aerobic.
2. Which cultures were the aerobes, the facultative anaerobes, and the anaerobes?

Figure 9.3 The appearance of aerobic and anaerobic growth in shake tubes.

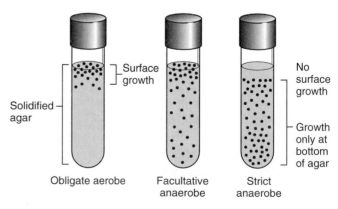

Obligate aerobe   Facultative anaerobe   Strict anaerobe

Figure 9.2 (a) Inoculating a melted agar deep. (b) Inoculating an agar slant.

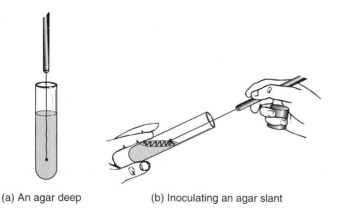

(a) An agar deep   (b) Inoculating an agar slant

EXERCISE

Laboratory Report: Aerobic and Anaerobic Growth

Results

Culture	A	B	C
Appearance of growth on slant			
Appearance and location of growth in deep			
Identity of culture A 　　　　culture B 　　　　culture C			

Questions

1. Would you expect an obligate anaerobe to grow on a slant incubated aerobically? Why?

2. Which kind of organism would you expect to grow both in the agar deep and on the slant?

3. Which kind of organisms can grow aerobically on slants?

4. Which of the three organisms you inoculated could grow throughout the agar deep and on top? (genus)

5. Why did you boil the agar deeps longer than it took to melt the agar?

6. If air can diffuse into agar and broth, how were the obligate anaerobes grown in the broth for the class? Your instructor will explain or demonstrate.

10

The Effect of Incubation Temperature on Generation Time

Getting Started

Every bacterial species has an optimal temperature—the particular temperature resulting in the fastest growth. Normally, the optimal temperature for each organism reflects the temperature of its environment. Organisms associated with animals usually grow fastest at about 37°C, the average body temperature of most warm-blooded animals. Organisms can divide more slowly at temperatures below their optimum, but there is a minimum temperature below which no growth occurs. Bacteria usually are inhibited at temperatures not much higher than their optimum temperature.

The effect of temperature can be carefully measured by determining the **generation time** at different temperatures. Generation time, or doubling time, is the time it takes for one organism to divide into two cells; on a larger scale, it is the time required for the population of cells to double. The shorter the generation time, the faster the growth rate.

Generation time can only be measured when the cells are dividing at a constant rate. To understand when this occurs, it is necessary to study the growth curve of organisms inoculated into a fresh broth medium. If plate counts are made of the growing culture, it can be seen that the culture proceeds through the four phases of growth: lag, log, stationary, and death (figure 10.1).

In the lag phase, the cells synthesize the necessary enzymes and other cellular components needed for growth. The cells then grow as rapidly as the conditions permit in the log phase, and when there are no longer sufficient nutrients or toxic product buildup, the cells go into the stationary phase. This is followed by the death phase. Only in the log phase are the cells growing at a constant maximum rate for the particular environment.

In this exercise, the generation time of *E. coli* will be compared when growing at two different temperatures. The growth of the cells can be mea-

Figure 10.1 Growth curve showing the four phases of growth.

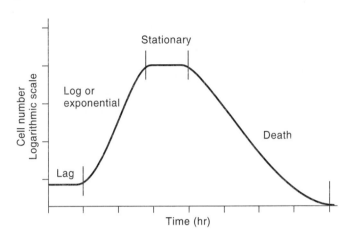

sured with a spectrophotometer or **colorimeter** because the number of cells in the culture is directly proportional to the absorbance (figure 10.2). That means that the **absorbance** (also called **optical density**) increases proportionately as the culture becomes increasingly more cloudy from the multiplication of the bacteria. Readings of the incubating cultures are taken every 20 minutes for 80 minutes. The results are then plotted, and the generation time is determined.

Definitions

Absorbance. A measure of turbidity.

Generation time. The time it takes for a population of cells to double.

Colorimeter. An instrument that can be used to measure the turbidity of bacterial growth.

Optical density (O.D.). An older, but still widely used, term for absorbance when used for measuring bacterial growth.

(a)

(b)

Objectives

1. To understand the phases of a growth curve.
2. To understand the effect of temperature on generation time.
3. To learn to calculate generation time.
4. To learn to use semi-log paper.

Reference

Nester et al. *Microbiology: A human perspective,* **4th ed., 2004. Chapter 4, Section 4.7.**

Materials

Per team

Cultures

Escherichia coli (TS broth cultures in log phase)

TS broth, 1, in a tube that can be read in a spectrophotometer or Klett colorimeter. Prewarmed in a water bath.

Water bath at 30°C (to be used by half the class)

Water bath at 37°C (to be used by the other half of the class)

Procedure

1. Add 0.5 ml–1.0 ml of an *E. coli* culture growing in log phase to 5.0 ml TS broth. The dilution is not important as long as the broth is turbid enough to be read at the low end of the scale (0.1 O.D. or a Klett reading of about 50). If you start with an O.D. that is too high, your last readings will reach the part of the scale that is not accurate (an O.D. of about 0.4 or about 200 on the Klett).
2. With a wavelength of 420, set the spectrophotometer at zero with an uninoculated tube of TS broth (which is termed a blank). Your instructor will give specific directions.
3. Take a reading of the culture and record it as 0 time. Return tube to the assigned water bath as quickly as possible because cooling slows the growth of the organisms. Wipe off water and fingerprints from the tubes before taking a reading.
4. Read the O.D. of the culture about every 20 minutes for about 80 minutes. Record the exact time of the reading so the data can be plotted correctly.
5. Record your data—the time and O.D. readings—in your manual and on the blackboard.

6. Plot the data on semi-log graph paper (page 81). Semi-log paper is designed to convert numbers or data to $\log_{10}$ as they are plotted on the y axis. The same results would be obtained by plotting the $\log_{10}$ of each of the data points on regular graph paper but semi-log paper simplifies this by permitting you to plot raw data and obtain the same line. Time is plotted on the horizontal x axis. Draw a straight bestfit line through the data points. The cells are growing logarithmically, so therefore the data should generate a straight line on log paper (figure 10.3).

7. Also plot the data from the other temperature by averaging the class data on the blackboard.

8. Calculate the generation time for *E. coli* at each temperature. This can be done by arbitrarily selecting a point on the line and noting the O.D. Find the point on the line where this number has doubled. The time between these two points is the generation time.

Figure 10.3 Growth curve of cells growing in log phase at 37°C and 30°C.

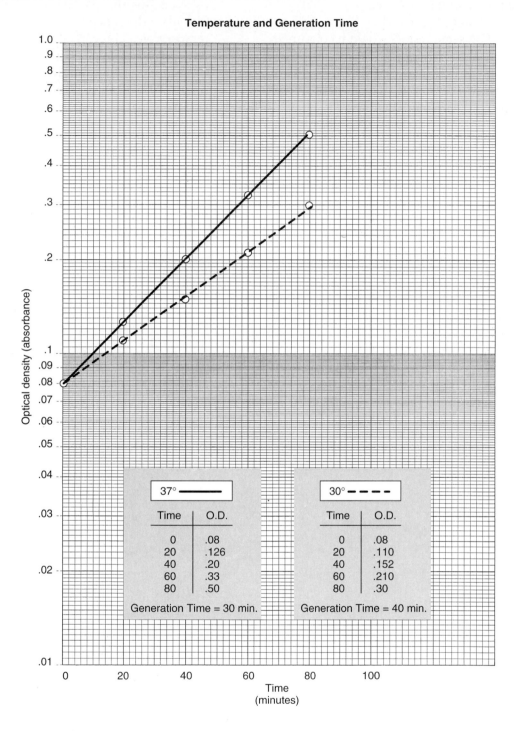

Temperature and Generation Time

37° ———			30° – – – –	
Time	O.D.		Time	O.D.
0	.08		0	.08
20	.126		20	.110
40	.20		40	.152
60	.33		60	.210
80	.50		80	.30
Generation Time = 30 min.			Generation Time = 40 min.	

EXERCISE

Laboratory Report: The Effect
of Incubation Temperature on Generation Time

Results

Data: Your Temperature _____   Class Average Temperature _____

	Time	Reading	Time	Reading
1				
2				
3				
4				
5				

Generation Time

E. coli at 37°C

E. coli at 30°C

Questions

1. What is the generation time of an organism?

2. Why is it important to keep the culture at the correct incubation temperature when measuring the generation time?

3. Why is it important to use cells in log phase?

4. If the growth of two cultures were plotted on semi-log paper, one slower than the other, which would have the steeper slope?

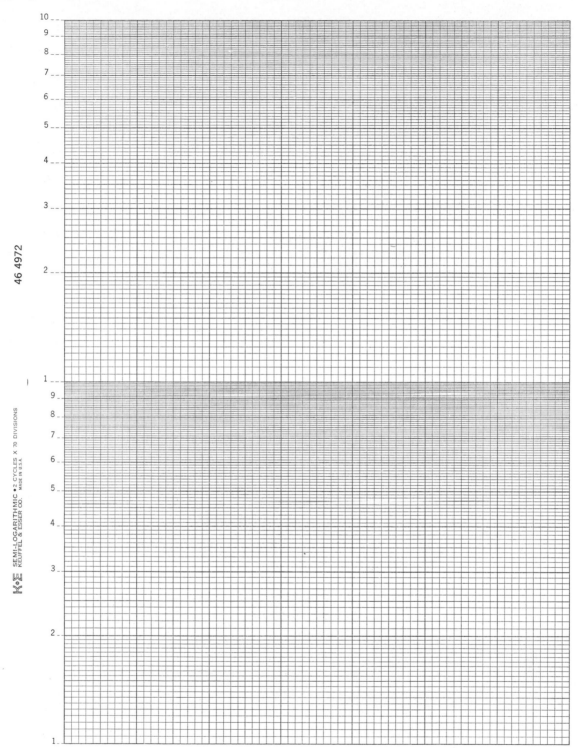

NOTES:

INTRODUCTION to Control of Microbial Growth

For many microbiologists, control of microbial growth means maximization of microbial growth, for example, when producing baker's yeast or in the production of antibiotics. To others such as physicians and allied members of the medical profession, control means minimization of microbial growth, for example, the use of heat and ultraviolet light to destroy microorganisms present in growth media, gloves, and clothing. It can also imply the use of **antiseptics, disinfectants,** and **antibiotics** to inhibit or destroy microorganisms present on external or internal body parts.

Historically, Louis Pasteur (1822–1895) contributed to both areas. In his early research, he discovered that beer and wine making entailed a fermentation process involving initial growth of yeast in the fermentation of liquor. Later, he showed that a sterile broth infusion in a swan-necked flask showed no turbidity due to microbial growth (figure I.5.1), and that upon tilting the flask, the sterile infusion became readily contaminated. The swan-necked flask experiment was both classical and monumental in that it helped resolve a debate, of more than 150 years, over the possible origin of microorganisms by **spontaneous generation** (abiogenesis).

The debate was finally squelched by John Tyndall, a physicist, who was able to establish an important fact overlooked by Pasteur—namely, that some bacteria in hay infusions existed in two forms: a **vegetative** form readily susceptible to death by boiling of the hay infusion, and a resting form now known as an **endospore,** which was resistant to boiling. With this knowledge, Tyndall developed a physical method of sterilization, which we now describe as **tyndallization,** whereby both vegetative cells and endospores are destroyed when the infusion is boiled intermittently with periods of cooling. For sterilization of some materials by tyndallization, temperatures below boiling are possible. Tyndallization, although a somewhat lengthy sterilization

---

Figure I.5.1 Pasteur's experiment with the swan-necked flask. (1–3) If the flask remains upright, no microbial growth occurs. (4 and 5) If microorganisms trapped in the neck reach the sterile liquid, they grow. From Eugene W. Nester et al., *Microbiology: A Human Perspective.* Copyright © 2003 The McGraw-Hill Companies. All Rights Reserved. Reprinted by permission.

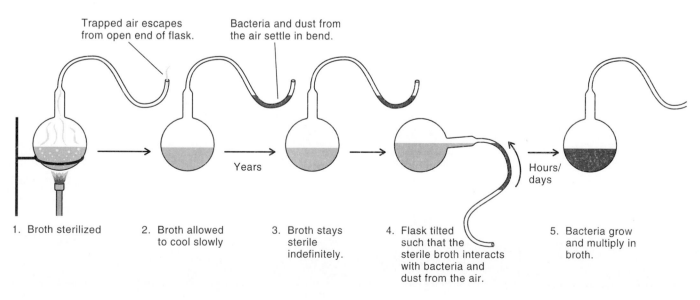

Trapped air escapes from open end of flask.

Bacteria and dust from the air settle in bend.

Years

Hours/ days

1. Broth sterilized
2. Broth allowed to cool slowly
3. Broth stays sterile indefinitely.
4. Flask tilted such that the sterile broth interacts with bacteria and dust from the air.
5. Bacteria grow and multiply in broth.

method, is sometimes used to sterilize chemical nutrients subject to decomposition by the higher temperatures of autoclaving.

At about this same time, chemical disinfectants for help in healing compound bone fractures were introduced by John Lister, an English surgeon, who was also impressed with Pasteur's findings. Lister had heard that carbolic acid (phenol) had remarkable effects when used to treat sewage in Carlisle; it not only prevented odors from farmlands irrigated with sewage, but also destroyed **entozoa,** intestinal parasites that usually infect cattle fed on such pastures.

The control of microbial growth has many applications today, both in microbiology and in such areas as plant and mammalian cell culture. Traditional examples of it include pure culture isolation, and preparation of sterile culture media, bandages, and instruments. It also includes commercial preparation of various microbial products such as antibiotics, fermented beverages, and food.

Exercises 7, 9, and 10 touch on maximization of microbial growth. In this section, the exercises deal with minimization or elimination of microbial growth by heat, ultraviolet light, osmotic pressure, antiseptics, and antibiotics.

… we are too much accustomed to attribute to a single cause that which is the product of several, and the majority of our controversies come from that.

Von Liebig

EXERCISE 11

Moist and Dry Heat Sterilization: Thermal Death Point and Thermal Death Time

Getting Started

Physical methods used in the hospital environment to control microbial growth include heat and ultraviolet light to kill microorganisms; filtration to remove microorganisms from growth media containing heat-labile substances such as enzymes, and from air in operating rooms and cell transfer rooms; and the use of sterile gloves, masks, and clothing in these rooms to help control air convection of microorganisms present on skin and hair.

In this exercise, some effects of heat sterilization are studied since heat is commonly used to sterilize many of the materials used in hospitals and laboratories. When heat is applied, most microbes are killed, whereas when cold temperatures are applied, inhibition of microbial growth is more likely to occur. The sensitivity of a microorganism to heat is affected by its environment and genetics. Environment includes factors such as incubation temperature, chemical composition of the growth medium, and the age and concentration of cells in the growth medium. Genetically, some microorganisms are more tolerant of heat than others. Examples include the ability of microorganisms, classified as **thermophiles,** to grow at higher temperatures than others, and the ability of some microorganisms to produce heat-resistant structures called endospores. Members of the genera *Bacillus* and *Clostridium* are capable of endospore production.

Heat is applied in either a dry or moist form. Examples of dry heat include hot air ovens used largely for sterilization of glassware such as petri dishes and pipets, and microincinerators used for sterilizing needles and loops. Dry heat kills by dehydrating microorganisms, which results in **irreversible denaturation** of essential enzyme systems. Sterilization with dry heat requires considerably more time and higher temperatures than with moist heat, because dry heat penetrates to the inside of microbial cells more slowly than does moist heat. Typical sterilization times and temperatures are 2 hours at 165°C for dry heat and 15 minutes at 121°C for moist heat. The mode of action is the same for both.

Autoclaving is the most commonly used method of moist heat sterilization. Some other moist heat methods are **pasteurization, boiling,** and **tyndallization.** With autoclaving and tyndallization, both the vegetative and endospore forms of microorganisms are killed, whereas with pasteurization and boiling, usually only vegetative cells are killed.

Pasteurization, which is named for Louis Pasteur, is a moist heat process used in beverages such as milk, beer, and wine to kill **pathogenic** bacteria and reduce the number of nonpathogenic bacteria such as **thermoduric** bacteria. The beverages are heated under controlled conditions of temperature and time, either 63°C for 30 minutes or 72°C for 15 seconds. However, many endospore-forming bacteria survive pasteurization. The lower temperatures of pasteurization help preserve food flavor.

Tyndallization, named after the physicist John Tyndall, is sometimes used to sterilize nutrient media subject to inactivation by the higher temperatures of conventional autoclaving. It is also useful in emergencies, such as when an autoclave becomes inoperative. Tyndallization is a lengthy process requiring three days. The solution to be sterilized is usually steamed for 30 minutes in flowing steam (100°C) on each of three successive days. Between steaming times, the solution is left at room temperature. In principle, the first boiling kills all vegetative cells, the second boiling destroys newly germinated endospores, and the third boiling serves as an added insurance that no living cells remain in the solution.

Boiling for 10 minutes is used to rid solutions such as drinking water of vegetative forms of pathogenic bacteria and other pathogens such as parasitic worms and protozoa.

Two methods for determining the heat sensitivity of a microorganism are the **thermal death point** (TDP) and the **thermal death time** (TDT). The TDP is defined as the lowest temperature necessary to kill all of the microorganisms present in a culture in 10 minutes. The TDT is defined as the minimal time necessary to kill all of the microorganisms present in a culture held at a given temperature. These general principles are commonly used when establishing sterility requirements for various processes. Examples include milk, food preservation, and hospital supplies.

Definitions

Antibiotic.   A chemical substance produced by a microorganism (a bacterium or a fungus) which has the capacity to inhibit the growth of or kill a disease producing microorganism.

Antiseptic.   A substance that prevents or arrests the growth or action of microorganisms, either by inhibiting their growth or by destroying them.

Autoclave.   A form of moist heat sterilization, conventionally performed at 121°C for 15 minutes.

Boiling.   Moist heat treatment that kills most pathogens. Conventionally, it is performed at 100°C; the time varies, although it is often done for 30 minutes.

Disinfectant.   An agent that frees from infection, *e.g.*, a chemical that destroys vegetative cells but ordinarily not bacterial endospores.

Dry heat oven.   Dry heat sterilization, conventionally done at 160°C to 170°C for 2–3 hours.

Irreversible denaturation.   A change in physical form of certain chemicals, which when heated, destroys them. Enzymes constitute one such example.

Pasteurization.   The use of moist heat at a temperature sufficiently high enough to kill pathogens but not necessarily all organisms. It is commonly used for milk-type products.

Pathogen.   An organism able to cause disease.

Thermal death point.   The lowest temperature able to kill all microbes in a culture after a given time.

Thermal death time.   The minimal time necessary to kill all microbes in a culture held at a given temperature.

Thermoduric.   Microbes able to survive conventional pasteurization, usually 63°C for 30 minutes or 72°C for 15 seconds.

Thermophile.   An organism able to grow at temperatures above 55°C.

Tyndallization.   The process of using repeated cycles of heating and incubation to kill spore-forming bacteria.

Objectives

1. To provide background information about physical sterilization methods requiring either moist heat or dry heat. Included in the discussion of moist heat methods are autoclaving, boiling, tyndallization, and pasteurization.
2. To introduce a quantitative laboratory method for determining the susceptibility of different bacteria to the lethal effect of moist heat—thermal death point (TDP) and thermal death time (TDT).
3. To demonstrate laboratory equipment commonly used for physical sterilization of moist and dry materials: the steam autoclave and the **dry heat oven.**

References

Block, S. S. (ed). *Disinfection, sterilization and preservation*, 4th ed. Philadelphia, PA: Lea & Febiger.

Frazier, W. C., and Westhoff, D. *Food microbiology*, revised. New York: McGraw-Hill Book Company, 1988. Contains relevant information on principles of food preservation and spoilage.

Nester et al. *Microbiology: A human perspective*, 4th ed., 2004. Chapter 5, Section 5.3.

Materials

Cultures

24-hour 37°C *Escherichia coli* cultures in 5 ml TS broth

Spore suspension in 5 ml of sterile distilled water of a 4–5 day 37°C nutrient agar slant culture of *Bacillus subtilis*

5 ml of TS broth, 5 tubes per student

Two large beakers or cans, one for use as a water bath and the other for use as a reservoir of boiling water (per student)

Either a ring stand with wire screen and a Bunsen burner or a hot plate (per student)

One thermometer (per student)

A community water bath

A vortex apparatus (if available)

Procedure

First Session: Determination of Thermal Death Point and Thermal Death Time

Note: A procedural culture and heating distribution scheme for 8 students is shown in table 11.1. For this scheme, each student will receive one broth culture, either *E. coli* or *B. subtilis*, to be heated at one of the assigned temperatures (40°C, 55°C, 80°C, or 100°C).

Note: As an alternative, instead of each student preparing his or her own water bath, the instructor can provide four community water baths preset at 40°C, 55°C, 80°C, and 100°C.

1. Suspend your assigned culture by gently rolling the tube between your hands, followed by aseptically transferring a loopful to a fresh tube of broth (label "0 time Control").

2. Fill the beaker to be used as a water bath approximately half full with water, sufficient to totally immerse the broth culture without dampening the test tube cap.
3. Place your tube of broth culture in the water bath along with an open tube of uninoculated broth in which a thermometer has been inserted for monitoring the water bath temperature.
4. Place the water bath and contents on either the ring stand or hot plate and heat almost to the assigned temperature. One or two degrees before the assigned temperature is reached, remove the water bath from the heat source and place it on your bench top. The temperature of the water bath can now be maintained by periodically stirring in small amounts of boiling water obtained from the community water bath.

   Note: Students with the 100°C assignment may wish to keep their water bath on the heat source, providing the water can be controlled at a low boil.

5. After 10 minutes of heating, resuspend the broth culture either by vortexing or by gently tapping the outside of tube. Aseptically transfer a loopful to a fresh tube of broth (label "10 minutes").
6. Repeat step 5 after 20, 30, and 40 minutes.
7. Write the initials of your culture as well your initials on all 5 tubes and incubate them in the 37°C incubator for 48 hours.

Demonstration of the Steam Autoclave and Hot Air Oven

The Steam-jacketed Autoclave

Note: As the instructor demonstrates the special features of the autoclave, follow the diagram in figure 11.1. Note the various control valves—their

Table 11.1 Culture and Heating Temperature Assignments (8 Students)

Bacterial Culture Assignment		Water Bath Temperature Assignment			
		40°C	55°C	80°C	100°C
Escherichia coli		1	2	3	4
Bacillus subtilis		5	6	7	8

Figure 11.1 Steam-jacketed autoclave. Entering steam displaces air downward and out through a port in the bottom of the chamber. Dry objects are placed in the autoclave in a position to avoid trapping air. Watery liquids generate their own steam.

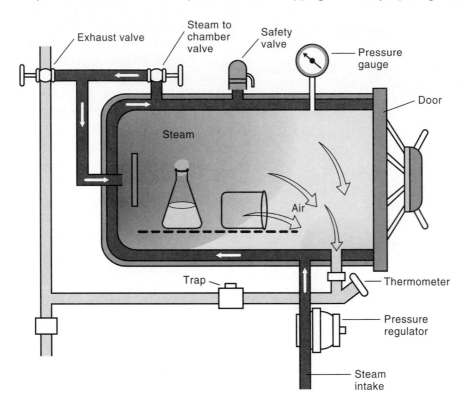

function and method of adjustment (exhaust valve, chamber valve, and safety valve); the steam pressure gauge; the thermometer and its location; and door to the chamber. Also make a note of special precautions necessary for proper sterilization.

1. Material sensitivity. Certain types of materials such as talcum powder, oil, or petroleum jelly cannot be steam sterilized because they are water repellent. Instead dry heat is used. Some materials are destroyed or changed by the standard autoclave temperature of 121°C; for example, some sugars are hydrolyzed and some medications are chemically changed. In such instances, the autoclave may be operated at a lower pressure and temperature for a longer period of time. A heat-sensitive fluid material can normally be sterilized by filtration. Filtration removes the bacteria. Keep in mind that the smaller the filter pores, the slower the rate of filtration.

2. Proper preparation of materials. The steam must be in direct contact with all materials to be sterilized. Therefore, media container closures such as metal caps with air passages, loosened screw cap lids, aluminum foil (heavy grade), and sometimes nonabsorbant cotton plugs are used.

3. Proper loading of supplies. There must be ample space between packs and containers so that the steam can circulate. When using cotton plugs, they should be loosely covered with foil to prevent moisture from the autoclave condensing on them during cooling.

4. Complete evacuation of air from the chamber. This is necessary before replacement with steam. Older models may require manual clearance, but in newer models it is automatic.

5. Proper temperature. Autoclaving at a pressure of 15 lb per square inch achieves a temperature of 121°C (250°F) at sea level. If

the temperature gauge does not register this reading, trapping of colder air in the autoclave is indicated, which lowers the temperature.

Note: The temperature is critical, not the pressure.

6. Adequate sterilization time. After the chamber temperature reaches 121°C, additional time is required for the heat to penetrate the material. The larger the size of individual containers and packs, the more time required. Time must be adjusted to the individual load size. Also, highly contaminated materials require more time.

7. Completion of the autoclaving process. Rapid reduction in steam pressure can cause fluids to boil vigorously through the caps or to explode. Drop steam pressure gradually as cooling occurs and, when possible, allow material to dry in the autoclave. If removed while moist, the wrappings or plugs may provide a means for reentry of bacteria present in the room air.

The Hot Air Oven

The hot air, or dry heat, oven is used in most laboratories for both drying glassware and sterilization.

When using the dry heat oven, the following guidelines are important:

1. Types of material suitable for sterilization include oil, petroleum jelly, Vaseline, metal containers, and dry, clean glassware.

2. An oven with circulating air takes about half the sterilization time of a static air oven. Better heat transfer occurs with circulating air.

3. Proper packaging is necessary to assure air circulation to the inside surfaces. For example, syringes must be separate from the plunger so that all surfaces are exposed to circulating air.

4. Sterilization of dirty materials should be avoided. The presence of extraneous materials such as protein delays the process and may allow bacteria to survive inside the material.

5. Heat-sensitive tapes designed for the autoclave cannot be used to assure adequate sterilization because the hot air oven requires much higher temperatures than does the autoclave. Bacterial spores, however, can be used.

6. In part 2 of the Laboratory Report, prepare a list of materials for your class that are sterilized in the autoclave. For each one, indicate the standard temperature, pressure, and time used for sterilization.

    Do the same for materials sterilized in the hot air oven. Indicate the oven sterilization temperature, time, and reasons for sterilizing there.

Second Session

After 48 hours, examine your broth tubes for the presence or absence of turbidity (growth). Write your results in the appropriate place in the table on the blackboard of your classroom. When all the results are entered, transfer them to table 11.2 of the Laboratory Report.

NOTES:

EXERCISE

11

Laboratory Report: Moist and Dry Heat Sterilization:
Thermal Death Point and Thermal Death Time

Results

1. Determination of thermal death point and thermal death time:

Table 11.2   Bacterial Growth at Assigned Temperatures and Times

Culture	40°C					55°C					80°C					100°C				
	C	10	20	30	40	C	10	20	30	40	C	10	20	30	40	C	10	20	30	40
E. coli																				
B. subtilis																				

Note: C = Control; 10, 20, 30, 40 = minutes of heating the inoculated culture at the assigned temperature; Use a + sign for growth and a – sign for no growth

a. Determine the thermal death time for each culture:

Thermal Death Time (Minutes)

Escherichia coli   ___

Bacillus subtilis   ___

b. Determine the thermal death point for each culture:

Thermal Death Point(°C)

Escherichia coli   ___

Bacillus subtilis   ___

2. Evaluation of materials sterilized in our laboratory with moist (autoclave) and dry heat (hot air oven):

    a. List of materials sterilized with the autoclave (see Procedure for criteria):

    b. List of materials sterilized with the hot air oven (see Procedure for criteria):

Questions

1. Discuss similarities and differences between determining thermal death point and thermal death time.

2. How would you set up an experiment to determine to the minute the TDT of *E. coli*? Begin with the data you have already collected.

3. A practical question related to thermal death time (TDT) relates to a serious outbreak of *E. coli* infection in early 1993 when people ate insufficiently grilled hamburgers. How would you set up an experiment to determine the TDT of a solid such as a hamburger? Assume that the thermal death point (TDP) is 67.2°C (157°F), the temperature required by many states to cook hamburger on an open grill. What factors would you consider in setting up such an experiment?

4. What is the most expedient method for sterilizing a heat-sensitive liquid that contains a spore-forming bacterium?

5. List one or more materials that are best sterilized by the following processes:

   a. Membrane filtration

   b. Ultraviolet light

   c. Dry heat

   d. Moist heat

   e. Tyndallization

   f. Radiation

6. What are three advantages of using metal caps rather than cotton for test tube closures? Are there any disadvantages?

7. How would you sterilize a heat-sensitive growth medium containing thermoduric bacteria?

8. Was the *Bacillus subtilis* culture sterilized after 40 minutes of boiling? If not, what is necessary to assure sterility by boiling?

EXERCISE

Control of Microbial Growth with Ultraviolet Light

Getting Started

Ultraviolet (UV) light is the component of sunlight that is responsible for sunburn. It can also kill microorganisms by acting on their DNA and causing mutations. It consists of very short **nonionizing wavelengths of radiation** (200 to 400 nm) located just below blue light (450 to 500 nm) in the **visible spectrum** (figure 12.1).

The actual mechanism of mutation is the formation of **thymine dimers** (figure 12.2). Two adjacent **thymines** on a DNA molecule bind to each other; when the DNA is replicated, an incorrect base pair is frequently incorporated into the newly synthesized strand. This may cause mutation and if there is sufficient radiation, ultimately, the death of the cell.

UV does not penetrate surfaces and will not go through ordinary plastic or glass. It is only useful for killing organisms *on* surfaces and in the air. Sometimes UV lights are turned on in operating rooms and other places where airborne bacterial contamination is a problem. Since UV light quickly damages the eyes, these lights are turned on only when no one is in the irradiated area.

Bacteria vary in their sensitivity to UV. In this exercise, the sensitivity of *Bacillus* endospores will

Figure 12.2 Thymine dimer formation. Covalent bonds form between adjacent thymine molecules on the same strand of DNA. This distorts the shape of the DNA and prevents replication of the changed DNA. From Eugene W. Nester et al., *Microbiology: A Human Perspective*. Copyright © 2003 The McGraw-Hill Companies. All Rights Reserved. Reprinted by permission.

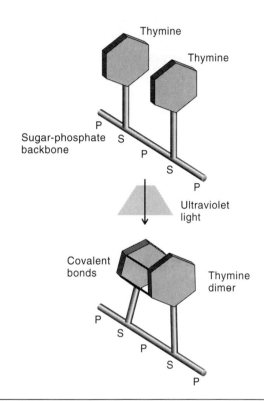

Figure 12.1 Germicidal activity of radiant energy.

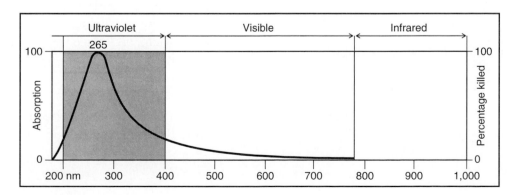

be compared with non-spore-forming cells. You will also irradiate a mixed culture such as organisms in soil or hamburger to compare the resistance of different organisms. Immediately following irradiation it is important to place the samples in a dark container since some cells potentially killed by UV recover when exposed to longer wavelengths of light (**light repair**).

Definitions

Light repair.   DNA repair in cells previously exposed to UV by a DNA repair enzyme that requires visible light.

Nonionizing wavelengths of radiation. Noncharged wavelengths of radiation with wavelengths between 175 nm and 350 nm. This light is invisible and is exemplified by UV light. Beta rays and X rays are examples of ionizing wavelengths of radiation.

Thymine dimer.   A molecule formed when two adjacent thymine molecules in the same strand of DNA covalently bond to one another.

Thymine.   One of the four nucleotide subunits of DNA.

Visible spectrum.   Uncharged wavelengths of radiation with wavelengths between 400 and 800 nm.

Reference

Nester et al. *Microbiology: A human perspective*, 4th ed., 2004. Chapter 5, Section 5.6.

Materials

Cultures

Suspension of *Bacillus* spores in sterile saline

E. coli in TS broth

Raw hamburger or soil mixed with sterile water

TS agar plates, 3 per team

UV lamp with shielding. An 18- to 36-inch fluorescent bulb is ideal. It enables uniform exposure of three to six partially opened petri dishes.

Sterile swabs, 3

Safety glasses for use with the UV lamp

Dark box for storing plates after UV exposure

Procedure

First Session

Safety Precautions: (1) The area for UV irradiation should be in an isolated part of the laboratory. (2) Students should wear safety glasses as a precautionary measure when working in this area. (3) Never look at the UV light after turning it on since it could result in severe eye damage. Skin damage is also a slight possibility.

1. Dip a sterile swab in a suspension of *Bacillus* spores and swab an agar plate in three directions as shown in figure 14.3.
2. Repeat the procedure with a suspended *E. coli* broth culture.
3. For the third plate, you can either dip the swab into a mixture of sterile water and hamburger or sterile water and soil.
4. Place the plates under a UV lamp propped up about 20 cm from the bench surface. Open the petri dish(es) and partially cover each plate with the lid. The part of the plate protected by the lid will be the control because UV does not penetrate most plastic.
5. Put on your safety glasses and turn on the UV light. Expose the plates to UV for 3 minutes.
6. Turn off the UV light. Cover the plates, invert them, and place them in a covered container. Incubate at 37°C for 48 hours.

Second Session

Observe the plates and record your findings in the Laboratory Report.

EXERCISE

Laboratory Report:
Control of Microbial Growth with Ultraviolet Light

Results

1. Record your observations for control and treated sides of petri dishes exposed to UV light at a distance of _____ cm for _____ minutes. Make a drawing of each plate.

   a. Plate containing *Bacillus* spores.

   b. Plate containing *E. coli.*

   c. Plate containing either a raw hamburger suspension or soil suspension. Indicate which one you used.

2. Which organisms were most resistant to UV? _____ Least resistant? _____

Questions

1. Why can't you use UV to sterilize microbiological media, e.g., agar or broth?

2. How does UV cause mutations?

3. Give a possible reason some organisms in the soil (or meat) were able to grow after exposure to UV but not others.

4. Frequently organisms isolated from the environment are pigmented, while organisms isolated from the intestine or other protected places are not. Can you provide an explanation for this?

5. Mutations can lead to cancer in animals. Explain why persons living in the southern half of the United States have a higher incidence of skin cancer than those in the northern half.

NOTES:

EXERCISE 13

Osmotic Pressure, and Its Effect on the Rate and Amount of Microbial Growth

Getting Started

Osmosis, which is derived from the Greek word "alter," refers to the process of flow or diffusion that takes place through a **semipermeable membrane.** In a living cell the cytoplasmic membrane, located adjacent to the inside of the cell wall, represents such a membrane (figure 13.1*a*).

Both the cytoplasmic membrane and the cell wall help prevent the cell from either bursting (**plasmoptysis,** figure 13.1*a*), or collapsing (**plasmolysis,** figure 13.1*b*) due to either entry or removal of water from the cell, respectively. The **solute** concentration both inside and outside the cell determines which, if any, process happens. When the solute concentration inside the cell is the same as the solute concentration on the outside of the cell (**isotonic**)*,* the cell remains intact. When the solute concentration outside the cell is less than the solute concentration inside the cell, an inward **osmotic pressure** occurs, and water enters the cell in an attempt to equalize the solute concentration on either side of the cytoplasmic membrane. If the solute concentration outside the cell is sufficiently low (**hypotonic**)*,* the cell will absorb water and sometimes burst (plasmoptysis). However, it rarely occurs due to the rigidity of the cell wall. The reverse phenomenon, cell shrinkage followed by cell lysis (plasmolysis), can occur when the cell is placed in a more concentrated (**hypertonic**) solution. This can become a life-and-death problem if too much water is removed from the cells (see figure 13.1).

When placed in an isotonic solution, some cells recover, although there are many genera that die once the external osmotic pressure exceeds their limitations. This concept is the basis of food preservation—the use of high salt concentrations (for cheese and pickle brine) and sugar concentrations (in honey and jams).

In general, fungi (yeasts and molds) are much more resistant to high external solute concentrations than are bacteria, which is one reason fungi

Figure 13.1 Movement of water into and out of cells. *(a)* Low and *(b)* very high salt-containing solutions. The cytoplasmic membrane is semipermeable and only allows water molecules to pass through freely.

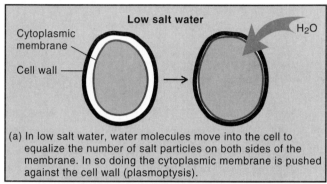

(a) In low salt water, water molecules move into the cell to equalize the number of salt particles on both sides of the membrane. In so doing the cytoplasmic membrane is pushed against the cell wall (plasmoptysis).

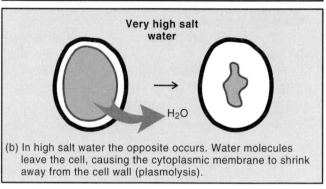

(b) In high salt water the opposite occurs. Water molecules leave the cell, causing the cytoplasmic membrane to shrink away from the cell wall (plasmolysis).

can grow in or on jelly, cheese, and fruit. There are, of course, exceptions among the bacteria; for example, the genus *Halobacterium* in the *Archaebacteria* is found in nature growing in water with a high salt content, e.g., Great Salt Lake in Utah; and the genus *Micrococcus halobius* in the family *Micrococcaceae* is sometimes found in nature growing on highly salted (25–32%) protein products such as fish and animal hides. It is interesting to note that all of these bacteria produce a red pigment.

There are also some salt-tolerant *Staphylococcus* strains able to grow at salt concentrations greater than 10% (w/v), allowing them to grow on skin surfaces. The salt-loving (**halophilic**) bacteria are also unique in that, like all *Archaebacteria*, they lack muramic acid as a bonding agent in their cell walls. Instead, their cell walls are believed to contain sodium and potassium ions. These ions help confer cell wall rigidity, perhaps helping explain the reason they require such high salt concentrations for good growth.

Yeasts and molds able to grow in high sugar (50–75%) and sometimes high salt concentrations (25–30%) are termed **saccharophilic** and **halophilic** fungi, respectively. Some of the yeast and fungus genera are *Debaromyces, Saccharomyces, Zygosaccharomyces, Aspergillus,* and *Penicillium.*

In this exercise, you examine the ability of some of the previously mentioned halophiles and saccharophiles to grow on the surface of TYEG agar plates containing increasing concentrations of salt and sugar. *Escherichia coli* is added as a salt-sensitive Gram-negative rod control. You will also have an opportunity to examine any changes in cell form with increasing salt and sugar concentrations.

It should be kept in mind that all of these halophilic and saccharophilic microbes are characterized by an increase in lag time and a decrease in growth rate, and in the amount of cell substance synthesized. In some ways, their **growth curve** (see exercise 10) parallels what happens when they are grown at a temperature below their optimal growth temperature. For example, halobacteria have a **generation time** of 7 hours and halococci of 15 hours.

Definitions

Generation time.   The time required for one cell to divide into two cells.

Growth curve.   A curve describing the four readily distinguishable phases of microbial growth: lag, log, stationary, and death.

Halophilic microbes.   A salt-requiring organism able to grow in a medium containing a salt concentration high enough to inhibit other organisms.

Hypertonic.   A fluid having an osmotic pressure greater than another fluid with which it is compared.

Hypotonic.   A fluid having an osmotic pressure lower than another fluid with which it is compared.

Isotonic.   A fluid having the same osmotic pressure as another fluid with which it is compared.

Osmotic pressure.   The pressure exerted by water on a membrane as a result of a difference in the concentration of solute molecules on each side of the membrane.

Plasmolysis.   Contraction or shrinking of the cytoplasmic membrane away from the cell wall due to a loss of water from the cell.

Plasmoptysis.   The bursting of protoplasm from a cell due to rupture of the cell wall when absorbing excess water from the external environment.

Saccharophilic microbes.   Microbes able to grow in environments containing high sugar concentrations.

Semipermeable membrane.   A membrane such as the cytoplasmic membrane of the cell which permits passage of some materials but not others. Passage usually depends on the size of the molecule.

Solute.   A dissolved substance in a solution.

Objectives

1. To provide an introduction to osmotic pressure and show how it may be used to inhibit growth of less osmotolerant microbes, while allowing more osmotolerant microbes to grow, although often at a considerably slower growth rate.
2. To show that some microorganisms either require or grow better in an environment containing high concentrations of salt (halophilic) or sugar (saccharophilic).

References

Mossel, D. A. A. "Ecological Essentials of Antimicrobial Food Preservation," pp. 177–195 in *Microbes and biological productivity.* Edited by D. E. Hughes and A. H. Rose. Cambridge University Press, 1971.

Nester et al. *Microbiology: A human perspective,* 4th ed., 2004. Chapter 3, Section 3.4, Chapter 4, Section 4.3, and Chapter 30, Section 30.2.

Materials

Cultures (may be shared by 2 to 4 students)

Use TYEG salts agar slants for *Escherichia coli*, *Micrococcus luteus*, and *Saccharomyces cerevisiae*

Use American Type Culture Collection (ATCC) medium 213 for the Preceptrol strain of *Halobacterium salinarium*

Incubate *E. coli*, *M. luteus*, and *S. cerevisiae* cultures for 24 hours at 35°C. Incubate *H. salinarium* culture for 1 week (perhaps longer) at 35°C

TYEG salts agar plates containing 0.5, 5, 10, and 20% NaCl, 4 plates

TYEG salts agar plates containing 0, 10, 25, and 50% sucrose, 4 plates

Procedure

First Session

Note: One student of the pair can inoculate the 4 TSA plates containing 0.5, 5, 10, and 20% NaCl, while the other student inoculates the 4 plates containing 0, 10, 25, and 50% sucrose.

1. First, use a glass-marking pencil to divide the undersurface of the 8 plates in quadrants and label with the initials of the four test organisms; for example, *E.c.* for *Escherichia coli*, etc. Also label the underside of each plate with the salt or sugar concentrations and your name.
2. Using aseptic technique, remove a loopful from a culture and streak the appropriate quadrant of your plate in a straight line approximately 1 inch long. Then reflame your loop, cool it for a few seconds, and make a series of cross streaks approximately one-half inch long in order to initiate single colonies for use in studying colonial morphology (figure 13.2). Repeat inoculation procedure for culture #1 in the appropriate quadrant of the remaining three agar plates.

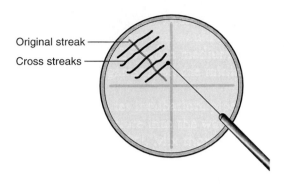

Figure 13.2
Streaking procedure for isolation of single colonies.

3. Repeat the inoculation procedure for the remaining three test organisms.
4. Invert and incubate the 8 plates at 30°C.
5. Observe the plates periodically (up to 1 week or more if necessary) for growth.

Second Session

1. Examine your plates for the presence (+) or absence (−) of growth. For growth, use 1 to 3 + signs (+ = minimal, + + = some, and + + + = good). Enter results in table 13.1 (various salt concentrations) and table 13.2 (various sugar concentrations) of the Laboratory Report.
2. Compare the colonial growth characteristics of cultures grown on agar media containing increasing salt and sugar concentrations. Make notes of any marked changes in colony color, colony size (in mm), and colony texture: dull or glistening, rough or smooth, and flat or raised. Record your findings in the Laboratory Report.
3. Prepare wet mounts of bacteria and yeast colonies showing marked changes in visual appearance with increasing salt and sugar concentrations. Examine bacteria with the oil immersion objective and yeasts with the high dry objective. Look for plasmolyzed cells and other changes such as cell form and size. Prepare drawings of any such changes in the Laboratory Report.

NOTES:

Name _____ Date _____ Section _____

EXERCISE

13

Laboratory Report: Osmotic Pressure, and Its Effect on the Rate and Amount of Microbial Growth

Results

1. Examination of petri dish cultures for the presence (+) or absence (−) of growth in the presence of increasing salt (table 13.1) and sugar concentrations (table 13.2). Use a series of one to three + signs to describe the amount of growth.

Table 13.1   Presence or Absence of Growth on TYEG Agar Plates Containing NaCl and Incubated for 48 Hours to 1 Week

Culture	NaCl Concentration (%)							
	0.5		5		10		20	
	48hr	1wk	48hr	1wk	48hr	1wk	48hr	1wk
Escherichia coli								
Halobacterium salinarium								
Micrococcus luteus								
Saccharomyces cerevisiae								

Table 13.2   Presence or Absence of Growth on TYEG Agar Plates Containing Sucrose and Incubated for 48 Hours to 1 Week

Culture	Sucrose Concentration (%)							
	0		10		25		50	
	48hr	1wk	48hr	1wk	48hr	1wk	48hr	1wk
Escherichia coli								
Halobacterium salinarium								
Micrococcus luteus								
Saccharomyces cerevisiae								

2. Comparison of the colonial growth characteristics of cultures inoculated on agar media containing increasing amounts of salt or sugar.

   a. *Escherichia coli*

NaCl %	Growth	Colony Color	Colony Size	Colony Texture
0.5				
5				
10				
20				

Sucrose %	Growth	Colony Color	Colony Size	Colony Texture
0				
10				
25				
50				

b. *Halobacterium salinarium*

NaCl %	Growth	Colony Color	Colony Size	Colony Texture
0.5				
5				
10				
20				

Sucrose %

0				
10				
25				
50				

c. *Micrococcus luteus*

NaCl %	Growth	Colony Color	Colony Size	Colony Texture
0.5				
5				
10				
20				

Sucrose %	Growth	Colony Color	Colony Size	Colony Texture
0				
10				
25				
50				

   d. *Saccharomyces cerevisiae*

NaCl %	Growth	Colony Color	Colony Size	Colony Texture
0.5				
5				
10				
20				

Sucrose %

Sucrose %				
0				
10				
25				
50				

3. Microscopic examination of wet mounts of bacteria and yeast colonies showing *marked* changes in visual appearance from the controls.

Questions

1. From your studies, which organism(s) tolerate salt best?_____ Least?_____
2. Which organism(s) tolerate sugar best?_____ Least?_____

3. Compare bacteria and yeast with respect to salt tolerance. Bear in mind both colonial and cellular appearance in formulating your answers.

4. Compare bacteria and yeast with respect to sugar tolerance. Bear in mind both colonial and cellular appearance in formulating your answer.

5. What evidence did you find of a *nutritional requirement* for salt or sugar in the growth medium?

6. Matching
   Each answer may be used one or more times.
   1. *Halobacterium*   ____ osmosensitive
   2. *Saccharomyces*   ____ long generation time
   3. *Escherichia coli*   ____ saccharophilic
   4. *Micrococcus*   ____ osmotolerant

7. Matching
   Choose the best answer. Each answer may be used one or more times, or not at all.
   1. Plasmolysis   ____ isotonic solution
   2. Plasmoptysis   ____ hypotonic solution
   3. Normal cell growth   ____ hypertonic solution
                              ____ swelling of cells

EXERCISE 14

Antiseptics and Antibiotics

Getting Started

It was during a visit through Central Europe in 1908 that I came across the fact that almost every farmhouse followed the practice of keeping a moldly loaf on one of the beams in the kitchen. When asked the reason for this I was told that this was an old custom and that when any member of the family received an injury such as a cut or bruise a thin slice from the outside of the loaf was cut off, mixed into a paste with water and applied to the wound with a bandage. I was assured that no infection would then result from such a cut.

Dr. A. E. Cliffe, a Montreal biochemist

Historically, chemicals such as antibiotics have been in existence a long time. However, the therapeutic properties of antibiotics simply were not recognized as such until Alexander Fleming's discoveries in the 1930s.

By definition, **antibiotics** are chemicals produced and secreted by microorganisms (bacteria, fungi, and actinomycetes) that can inhibit or destroy the growth of **pathogenic** microorganisms, often by altering an **essential metabolic pathway.** To be effective in human medicine, ideally they should be non-toxic to the human host and should discourage the formation of microbial strains resistant to the antibiotic (see exercise 15). Many antibiotics in use today are synthesized in the chemical laboratory.

Another interesting group of synthetic chemicals that act as antimetabolites are sulfa drugs, which originate from the azo group of dyes. One of these drugs, marketed under the name of Prontosil, inhibited microbial growth when tested *in vivo* (initially in mice and later in man), but when tested ***in vitro*** (test tubes) against streptococci nothing happened. In the 1930s, scientists at the Pasteur Institute in Paris showed that if Prontosil was chemically reduced it was not only active in the body, but

Figure 14.1 *(a)* Structures of sulfanilamide (sulfa drug) and of para-aminobenzoic acid (PABA). The portions of the molecules that differ from each other are shaded. *(b)* Reversible competitive inhibition of folic acid synthesis by sulfa drug. The higher the concentration of sulfa drug molecules relative to PABA, the more likely that the enzyme will bind to the sulfa drug, and the greater the inhibition of folic acid synthesis. From Eugene W. Nester et al., *Microbiology: A Human Perspective.* Copyright © 2004 The McGraw-Hill Companies. All Rights Reserved. Reprinted by permission.

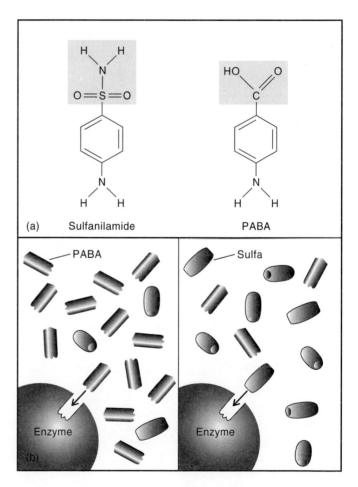

(a) Sulfanilamide   PABA

(b)

also able to kill bacteria *in vitro*. Prontosil was active *in vivo*, (but not *in vitro*) because enzymes within mice and men reduce Prontosil to a smaller active molecule, known as sulfanilamide (figure 14.1). The inhibitory action of the sulfanilamide-type drugs is one of **competitive inhibition,** in which the sulfanilamide acts as an **antimetabolite.** The sulfanilamide component replaces para-aminobenzoic acid (PABA), an **essential metabolite,** as a necessary part of folic acid, an essential **coenzyme** involved in amino acid synthesis.

Antiseptics are chemicals able to inhibit *in vivo* **sepsis** (infection). They do not need to kill the sepsis-producing agent, merely inhibit its growth. Antiseptic chemicals must be sufficiently nontoxic to allow application to skin and mucous membranes, such as the use of Listerine® for gargling.

These same chemicals act as **disinfectants** (a chemical able to kill vegetative forms but not necessarily spore forms of bacteria) when used at higher concentration levels. Toxicity is a major factor in determining usage of a chemical as either an antiseptic or disinfectant. Most *inorganic* heavy metal compounds can be used only as disinfectants (for example, mercurials). *Organic* heavy metal mercurial compounds (like mercurochrome) and also quaternary ammonium compounds and phenolics can be used as either antiseptics or disinfectants, depending on the concentration used.

The mode of action of antiseptics and disinfectants differs from antibiotics in that they act largely by denaturing proteins. They also lower surface tension, which is accompanied by cytoplasmic membrane dissolution (quaternary ammonium compounds), and act as oxidizing agents (chlorine-containing chemicals). A list of commonly used antiseptics and disinfectants and their area of application is shown in table 14.1.

An assay still used in many clinical laboratories to test the potency of antibiotics and drugs is a filter paper disc-agar diffusion procedure commonly known as the Kirby-Bauer test. A filter paper disc-agar diffusion method is also used for determining the potency of antiseptics. In this exercise, you will

Table 14.1 Chemical Compounds Commonly Used in Hospitals for Controlling Growth of Microorganisms

Sodium hypochlorite (5%)	Disinfectant	External surfaces, such as tables
Iodine (1% in 70% alcohol)	Disinfectant	External surfaces, such as tables
Iodophors (70 ppm avail. I₂)	Disinfectant	External surfaces, such as tables
Lysol (5%), a solution of cresol with soap	Disinfectant	External surfaces, such as tables
Phenol (5%), carbolic acid, source coal tar	Disinfectant	External surfaces, such as tables
Hexachlorophene (pHisoHex, Dial soap)	Disinfectant	Presurgical hand washing
Formaldehyde (4%)	Disinfectant	Oral and rectal thermometers
Iodophors (70 ppm avail. I₂)	Disinfectant	Oral and rectal thermometers
Zephrin (.001%)	Disinfectant	Oral and rectal thermometers
Alcohol, ethanol (70%)	Antiseptic	Skin
Iodine (tincture in alcohol with KI)	Antiseptic	Skin
Iodophors	Antiseptic	Skin
Organic mercury compounds (merthiolate, mercurochrome)	Antiseptic	Skin
Hydrogen peroxide (3%)	Antiseptic	Superficial skin infections
Potassium permanganate	Antiseptic	Urethral, superficial skin fungus infections
Silver nitrate (1%) (Argyrol)	Antiseptic	Prevention of eye infections in newborn babies
Zinc oxide paste	Antiseptic	Diaper rash
Zinc salts of fatty acids (Desenex)	Antiseptic	Treatment of athlete's foot
Glycerol (50%)	Antiseptic	Prevent bacterial growth in stool and surgical specimens
Ethylene oxide gas (12%)	Sterilization	Linens, syringes, etc.
Formaldehyde (20% in 70% alcohol)	Sterilization	Metal instruments
Glutaraldehyde (pH 7.5 or more)	Sterilization	Metal instruments

have an opportunity to determine both antiseptic and antibiotic potency with a modified Kirby-Bauer test.

Discs of filter paper impregnated with antibiotic solutions, in the same range of concentrations obtainable in the human body, are placed on an agar plate heavily seeded with the test bacterium. When incubated, the bacteria grow in a smooth lawn of confluent growth except in a clear zone around the antibiotic disc. The clear growth inhibition zone does not necessarily indicate degree of microbial susceptibility to the antibiotic; for zone size itself does not indicate if the antibiotic is appropriate for use in clinical treatment. When conducting the precisely controlled Kirby-Bauer test, various special conditions such as 2 to 5 hour cultures, controlled inoculum size, and short incubation periods are necessary. Such conditions are difficult to achieve in the time frame of most classrooms. Consequently, the simpler test performed in this exercise will demonstrate the Kirby-Bauer test principle without the added restrictions. Upon observation, the diameter of the clear zone of inhibition surrounding each antibiotic disc can be compared to that contained for the antibiotic on a standard chart (table 14.2). From this chart, one can determine if the test bacterium is resistant, intermediate, or sensitive to the antibiotic in question. A comparison chart is not available for antiseptics.

Table 14.2   Chart Containing Zone Diameter Interpretive Standards for Determining the Sensitivity of Bacteria to Antimicrobial Agents

Antimicrobial Agent	Disc Content	Zone Diameter, Nearest Whole mm		
		Resistant	Intermediate	Susceptible
Penicillin				
when testing staphylococci[b,c]	10 units	≤28	–	≥29
when testing enterococci[d]	10 units	≤14	–	≥15[d]
when testing streptococci[e]	10 units	≤19	20–27	≥28
Streptomycin				
when testing enterococci for high level resistance	**300 μg**	**6**	**7–9[q]**	**≥10**
when testing other organisms	**10 μg**	**≤11**	**12–14**	**≥15**
Tetracycline[m]	30 μg	≤14	15–18	≥19
Chloramphenicol	30 μg	≤12	13–17	≥18
Sulfonamides[n,o]	300 μg	≤12	13–16	≥17
Erythromycin	15 μg	≤13	14–22	≥23

National Committee for Clinical Laboratory Standards. *Performance Standards for Antimicrobial Disk Susceptibility Tests,* Fifth Edition; Approved Standard. Copyright © 1993 The National Committee for Clinical Laboratory Standards. By permission.
NOTE: Information in boldface type is considered tentative for one year.
[b]Resistant strains of *Staphylococcus aureus* produce β-lactamase and the testing of the 10-unit penicillin disc is preferred. Penicillin should be used to test the susceptibility of all penicillinase-sensitive penicillins, such as ampicillin, amoxicillin, azlocillin, bacampicillin, hetacillin, carbenicillin, mezlocillin, piperacillin, and ticarcillin. Results may also be applied to phenoxymethyl penicillin or phenethicillin.
[c]Staphylococci exhibiting resistance to methicillin, oxacillin, or nafcillin should be reported as also resistant to other penicillins, cephalosporins, carbacephems, carbapenems, and β-lactamase inhibitor combinations despite apparent *in vitro* susceptibility of some strains to the latter agents. This is because infections with methicillin-resistant staphylococci have not responded favorably to therapy with β-lactam antibiotics.
[d] The "Susceptible" category for penicillin or ampicillin implies the need for high-dose therapy for serious enterococcal infections. If possible, this should be denoted by a footnote on the susceptibility report form. Enterococcal endocarditis requires combined therapy with high-dose penicillin or high-dose ampicillin, or vancomycin, or teicoplanin plus gentamicin or streptomycin for bactericidal action. Since ampicillin or penicillin resistance among enterococci due to β-lactamase production is not reliably detected using routine disc or dilution methods, a direct, nitrocefin-based β-lactamase test is recommended. Synergy between ampicillin, penicillin, or vancomycin and an aminoglycoside can be predicted for enterococci by using a high-level aminoglycoside screening test.
[e]A penicillin MIC should be determined on isolates of *viridans Streptococcus* from patients with infective endocarditis.
[m]Tetracycline is the class disc for all tetracyclines, and the results can be applied to chlortetracycline, demeclocycline, doxycycline, methacycline, minocycline, and oxytetracycline. However, certain organisms may be more susceptible to doxycycline and minocycline than to tetracycline (such as some staphylococci and *Acinetobacter*).
[n]Susceptibility data for cinoxacin, nalidixic acid, nitrofurantoin, norfloxacin, sulfonamides, and trimethoprim apply only to organisms isolated from urinary tract infections.
[o]The sulfisoxazole disc can be used to represent any of the currently available sulfonamide preparations. Blood-containing media except for lysed horse blood are generally not suitable for testing sulfonamides or trimethoprim. Mueller-Hinton agar should be checked for excessive levels of thymidine as described in table 3.
[q]If the zone is 7 to 9 mm, the test is inconclusive and an agar dilution or broth microdilution screen test should be performed to confirm resistance.

Definitions

Antibiotic. A chemical produced largely by certain bacteria and fungi that can inhibit or destroy the growth of other organisms including pathogenic microorganisms.

Antimetabolite. A substance that inhibits the utilization of a metabolite necessary for growth (see figure 14.1).

Antiseptic. A chemical that inhibits or kills microbes. The definition also implies that the chemical is sufficiently nontoxic that it may be applied to skin and mucous membranes.

Coenzyme. Any heat-stable, nonprotein compound that forms an active portion of an enzyme system after combination with an enzyme precursor (apoenzyme). Many of the B vitamins are coenzymes.

Competitive inhibition. The inhibition of enzyme activity caused by the competition between the inhibitor and the substrate for the active (catalytic) site on the enzyme.

Disinfectant. A chemical agent that rids an area of pathogenic microorganisms. In so doing, it kills vegetative forms of bacteria but ordinarily not spore forms. The definition also implies that the chemical is sufficiently toxic that it should not be applied to body surfaces, only to material objects.

Essential metabolic pathway. A pathway of chemical transformations necessary for growth; if inhibited, the organism usually dies. The Krebs cycle and Embden-Meyerhof pathway are classic examples.

Essential metabolite. A chemical necessary for proper growth.

In vitro. Growth of microbes in test tubes.

In vivo. Growth of microbes in live plants and animals.

Pathogen. Any agent capable of causing disease, usually a microorganism.

Sepsis. The presence of pathogenic microorganisms or their toxins in tissue or blood.

Objectives

1. To provide introductory information about the origin and usage of antibiotics, antiseptics, and disinfectants.
2. To provide you with an opportunity to evaluate the bacteriostatic activity of antibiotics, antiseptics, and disinfectants with a modified Kirby-Bauer test.

Reference

Nester et al. *Microbiology: A human perspective,* 4th ed., 2004. Chapter 5, Section 5.1 and Chapter 21, Section 21.4.

Materials

Cultures (per team of 2 to 4 students)

Bacteria (24 hour 37°C TS broth cultures)

Staphylococcus epidermidis (a Gram-positive coccus)

Escherichia coli (a Gram-negative rod)

Pseudomonas aeruginosa (nonfermenting Gram-negative rod)

Mycobacterium smegmatis (acid-fast rod)

Vials of the following antibiotic discs, or dispensers:

Penicillin, 10 μg; streptomycin, 10 μg; tetracycline, 30 μg; chloramphenicol, 30 μg; nystatin, 100 units; sulfanilamide (or another sulfonamide), 300 μg; erythromycin, 15 μg

Beakers containing 10 ml aliquots of the following chemicals: 70% ethanol, 3% hydrogen peroxide, antiseptic mouthwash such as Listerine®, and 0.3% hexachlorophene

Mueller-Hinton agar, 6 plates

Sterile cotton swabs, 6

Sterile filter paper discs, ¼″ or ½″, 8

Small forceps, 1 per student

Ruler divided in mm

Procedure

Filter Paper Disc Technique for Antiseptics and Disinfectants: First Session

1. With a glass-marking pencil, divide the underside of two plates of Mueller-Hinton agar into quadrants and label them 1 through 4.
2. Record codes for the four antiseptics and disinfectants on the bottom sides of the two agar plates, one code for each quadrant: 70% ethanol: E, 3% hydrogen peroxide: HP, Listerine: L, and hexachlorophene: H.
3. Label the cover of one petri dish *S. epidermidis* and the cover of the other dish *E. coli*.
4. Suspend the *S. epidermidis* culture, then insert and moisten a sterile swab, remove excess, followed by streaking the swab in all directions on the surface of the agar plate. Discard swab in the appropriate waste container.
5. Repeat step 4 with *E. coli*.
6. Sterilize forceps by dipping them in 95% alcohol and then touch to the flame of the Bunsen burner. Air cool.
7. Using forceps, remove one of the filter paper discs from the container and dip it into solution 1: 70% ethanol.
8. Drain the disc thoroughly on a piece of clean absorbent toweling and place it in the center of quadrant 1 of the dish labeled *S. epidermidis* (figure 14.2). Tap disc gently.
9. Repeat steps 5, 6, and 7 and place the disc in the center of quadrant 1 of the plate labeled *E. coli*.
10. Repeat steps 5 through 8 for the remaining three compounds, using first 3% hydrogen peroxide, then the antiseptic mouthwash, and last hexachlorophene.
11. Invert the petri dishes and incubate at 37°C for 48 hours.

Filter Paper Disc Technique for Antibiotics: First Session

1. Divide the four broth cultures among team members, so that each student sets up at least one susceptibility test.
2. With a glass-marking pencil, divide the underside of four plates of Mueller-Hinton agar into six pie-shaped sections (figure 14.3a).

Figure 14.2(a–d) Filter paper disc technique for antiseptics.

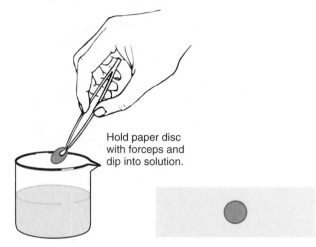

(a) Antiseptic solution

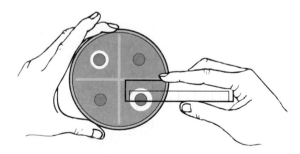

Hold paper disc with forceps and dip into solution.

(b) Drain disc on toweling.

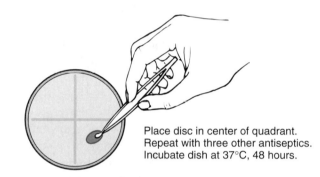

Place disc in center of quadrant. Repeat with three other antiseptics. Incubate dish at 37°C, 48 hours.

(c) Petri dish seeded with *S. aureus* or *E. coli*

(d) Measure the clear zone of inhibition surrounding each disc.

3. Record the codes of the seven antibiotic discs on the bottom side of the four plates, one code for each section with the remaining code for nystatin in the center of the plate. See table 14.4 of the Laboratory Report for code designations.
4. Label the cover of each plate with the name of the respective bacterium (see Materials for names).

Figure 14.3 Antibiotic susceptibility test. *(a)* The underside of a Mueller-Hinton agar plate showing the marking of sections and the arrangement for placement of antibiotic discs on the agar surface. *(b)* Procedure for streaking an agar plate in three or more directions with a swab inoculum in order to achieve a uniform lawn of growth.

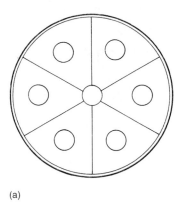

(a)

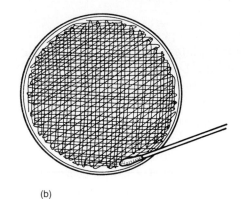

(b)

5. Using aseptic technique, streak the first broth culture as shown in figure 14.3*b*. The remaining three cultures should be streaked on separate plates in a similar manner.

6. Heat sterilize forceps (see step 6 of previous Procedure), and remove an antibiotic disc from the container. Place gently, with identification side up, in the center of one of the pie-shaped sections of the agar plate (see figure 14.3*a*). Tap gently to fix in position.

7. Continue placing the remaining six antibiotic discs in the same way, placing the last disc in the center where the lines cross.

   Note: Be sure to flame the forceps with alcohol after placing each disc because it is possible to contaminate stock vials with resistant organisms.

   Note: If a disc dispenser is used, follow the manufacturer's instructions.

8. Repeat steps 5 through 7 with the remaining three cultures.

9. Invert and incubate the plates at 37°C for 48 hours.

Filter Paper Disc Technique for Antiseptics and Disinfectants: Second Session

1. Turn over the *S. epidermidis* plate and with a ruler calibrated in mm, determine the diameter of the clear zone surrounding each disc. Repeat with the *E. coli* plate.

   Note: It may be necessary to illuminate the plate in order to define the clear zone boundary.

2. Record your results in table 14.3 of the Laboratory Report.

Filter Paper Disc Technique for Antibiotics: Second Session

1. Observe plates using the same method described in step 1 of the second session for antiseptics and disinfectants. In addition, make note of any large colonies present in the clear zone of growth inhibition surrounding each antibiotic disc. They may be resistant mutants.

2. Record your findings in table 14.4 of the Laboratory Report.

3. Compare your results where possible with table 14.2 and indicate in table 14.4 the susceptibility of your test cultures (*when possible*) to the antibiotics as resistant (R), intermediate (I), or susceptible (S).

Note: Your answers may not agree exactly with those in table 14.2 since this is a modified Kirby-Bauer test.

EXERCISE

14

Laboratory Report: Antiseptics and Antibiotics

Results

1. Filter paper disc technique for antiseptics and disinfectants:

Table 14.3 Bacteriostatic Activity of Various Antiseptics and Disinfectants

Antiseptic or Disinfectant	Zone of Inhibition (mm)	
	Staphylococcus epidermis	*Escherichia coli*
70% ethanol (E)		
3% hydrogen peroxide (HP)		
Listerine® (L)		
0.3% hexachlorophene (H)		
Others:		

What general conclusions can you make from this study? What differences, if any, did you observe on your plates between antiseptic and disinfectant preparations?

2. Filter paper disc technique for antibiotics:

Table 14.4   Antibiotic Susceptibility (Modified Kirby-Bauer Test)

Test Organism	Zone of Inhibition (mm)							Susceptibility[a]							Notes
	[b]Chl	Ery	Pen	Str	Sul	Tet	Nys	[b]Chl	Ery	Pen	Str	Sul	Tet	Nys	
S. epidermidis															
E. coli															
P. aeruginosa															
M. smegmatis															

[a]R = Resistant, I = Intermediate, and S = Susceptible.
[b]Chl = Chloramphenicol (Chloromycetin), 30 µg; Ery = Erythromycin, 15 µg; Pen = Penicillin G, 10 µg; Str = Streptomycin, 10 µg; Sul = Sulfanilamide, 300 µg; Tet = Tetracycline, 30 µg; Nys = Nystatin, 100 units.

Questions

1. What relationship did you find, if any, between the Gram-staining reaction of a microorganism and its susceptibility to antiseptics and disinfectants?

2. You may have noted that nystatin was not listed in table 14.2. The reason for its omission is that it is an antifungal antibiotic. Was it antibiotically active against any of the bacteria you studied? Is there an organism that could have been tested that might have been susceptible?

3. Name some other factors affecting the size of the zone of growth inhibition that were not included in your modified Kirby-Bauer test. Why were they omitted? Discuss their importance. The *Manual of Clinical Microbiology*, 5th edition, eds. Balows, Hauser, Herrman, Isenberg, and Shadomy, published by the American Society for Microbiology, Washington, D.C., 1991, is an excellent reference.

4. To what general groups of organic compounds does hexachlorophene belong? What are the advantages and disadvantages of using hexachlorophene for surgical scrub-downs?

5. Matching
   1. 70% ethanol   ___ antibiotic
   2. 5% phenol   ___ antiseptic
   3. nystatin   ___ coal tar dye
   4. prontosil   ___ disinfectant
   5. sulfanilamide   ___ drug

NOTES:

Plate 1 Appearance of bacterial colonies growing on agar. © Larry Jensen/Visuals Unlimited.

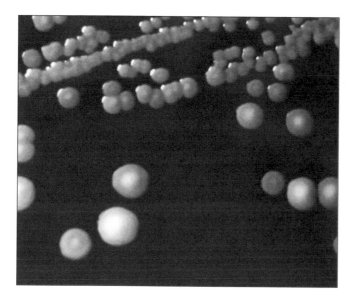

Plate 2 Appearance of mold (*Penicillium*) growing on an agar plate. © Raymond B. Otero/Visuals Unlimited.

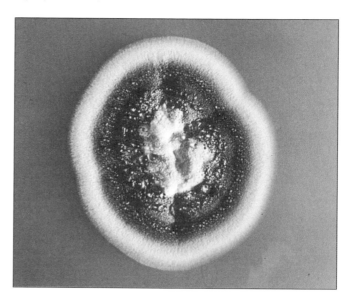

Plate 3 Bacterial shape. *Staphylococcus aureus* cocci as seen with the bright-field light microscope (×1,000). © LeBeau/Biological Photo Service.

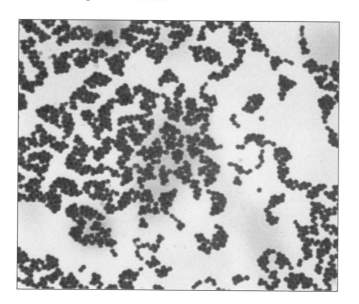

Plate 4 Bacterial shape. *Bacillus megaterium* rods as seen with the bright-field light microscope (×600). © George Wilder/Visuals Unlimited.

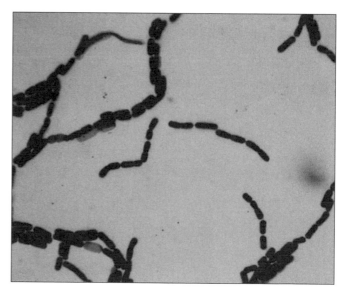

Plate 5 Bacterial shape. *Rhodospirillum rubrum* as seen with the bright-field light microscope (×500). © Thomas Tottelben/Tottelben Scientific Co.

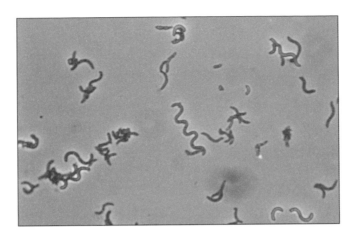

Plate 6 Budding yeast photographed with bright-field (*left*) and dark-field (*right*) microscopy. © Dr. Edward Bottone, The Mount Sinai Hospital, New York, New York.

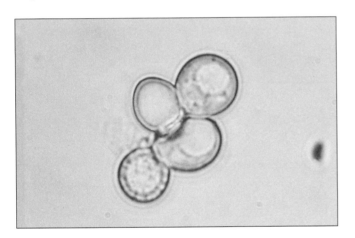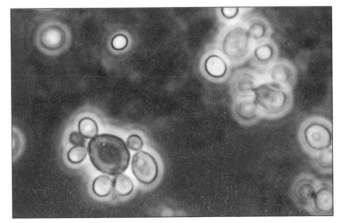

Plate 7 Light micrograph (×900) of a Gram-stained mixture of Gram-positive *Staphylococcus aureus* (purple cocci) and Gram-negative *Escherichia coli* (pink rods). Courtesy of John Harley.

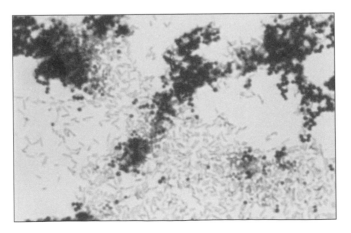

Plate 8   Macroscopic fungi. (*a*) *Calvatia gigantea,* one of the largest visible fungi, which is estimated to produce as many as 26 million reproductive spores; (*b*) *Polyporus arcularius,* a large visible bracket fungus growing on a decaying tree trunk. (*a*) © Glenn M. Oliver/Visuals Unlimited (*b*) ©Dick Poe/Visuals Unlimited.

(a)

(b)

Plate 9   Green mold (probably a *Penicillium* sp.) prevalent on stored fruits. The white areas consist largely of fungus mycelium, which removes nutrients from the strawberry, and eventually forms asexual reproductive structures containing spores (conidia), which when released by air currents further spread the infection (green area on right bottom of strawberry). © Matt Meadows/Peter Arnold, Inc.

Plate 10   Ringworm lesions on the scalp. Note hair loss and scaling of scalp. © Everett S. Beneke/Visuals Unlimited.

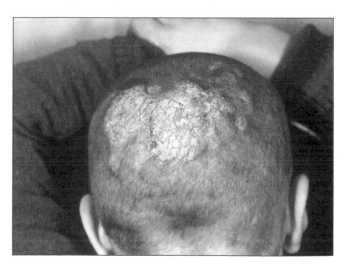

Plate 11   Fixed cutaneous sporotrichosis. Identification is based on the type of lesion formed and culture characteristics (see also color plates 12, 13, and 14). © Everett S. Beneke/Visuals Unlimited.

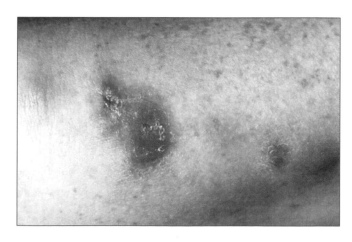

Plates 12, 13, and 14 *Sporotrichum schenkii,* a dimorphic fungus that forms a moldlike colony with hyphae containing terminal clusters of pyriform conidia and mycelia when incubated at 20°C (plates 12 and 13) and fusiform to round yeast cells when incubated at 37°C (plate 14). Courtesy of the Upjohn Co.

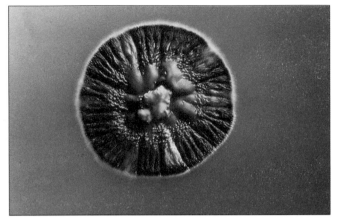

Plates 15, 16, and 17 *Coccidioides immitis,* a dimorphic fungus that forms a white, moldlike colony containing septate hyphae and chains of thick-walled arthrospores when grown on an agar medium at 20° to 37°C (plates 15 and 16), and round, thick-walled spherules (20–80 μm in diameter) containing many small endospores (2–5 μm in diameter) when isolated from sputum, pus, gastric contents, or spinal fluid (plate 17). Courtesy of the Upjohn Co.

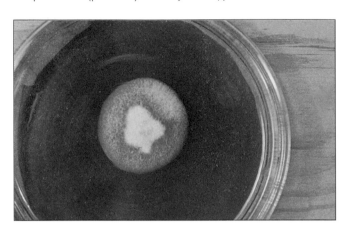

Plate 13

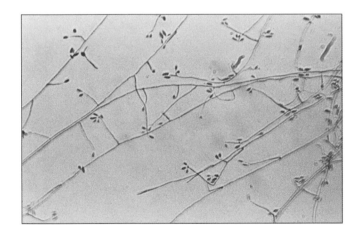

Plate 16

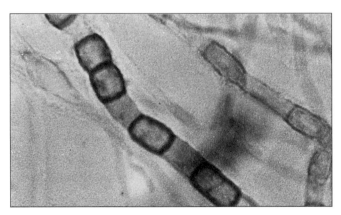

Plate 14

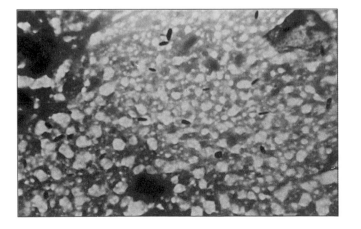

Plate 17

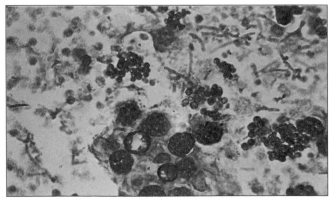

Plate 18   *Paramecium* from pond water showing cilia and internal structures. Phase-contrast microscopy (×100). © Mike Abbey/Visuals Unlimited.

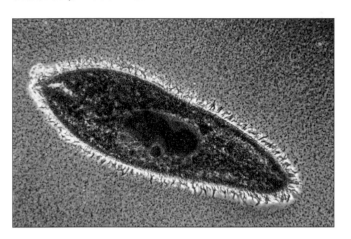

Plate 19   *Schistosoma miracidium,* or snail phase. Note the ciliated larvae.© Cabisco/Visuals Unlimited.

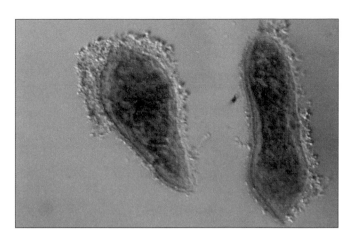

Plate 20   Fermentation results. From left (yellow) to right: 1. Uninoculated control. 2. No change. 3. Acid and gas. 4. Acid. Courtesy of the University of Washington.

Plate 21   Indol test. From left to right: 1. Uninoculated control. 2. Positive for indol formation (red). 3. Negative for indol formation. Courtesy of the University of Washington.

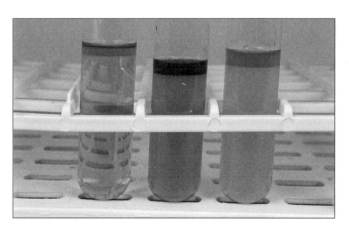

Plate 22   Citrate utilization. From left to right: 1. Uninoculated control. 2. No growth citrate negative. 3. Growth citrate positive. Courtesy of the University of Washington.

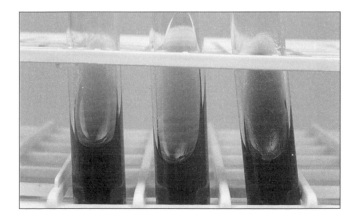

Plate 23   Urease test. From left to right: 1. Uninoculated control. 2. Urease negative. 3. Urease positive. Courtesy of the University of Washington.

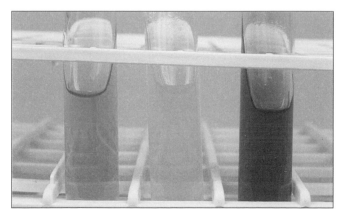

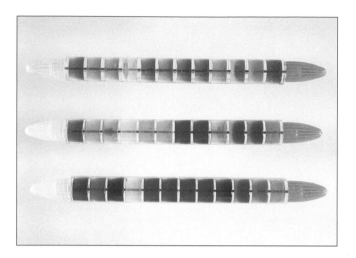

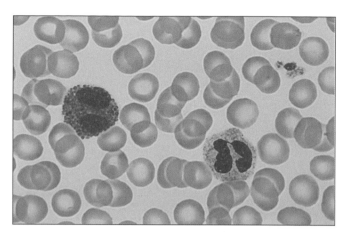

Plate 26 Note line of identity by fusion. See figure 31.3*a* for further explanation. Courtesy of the University of Washington.

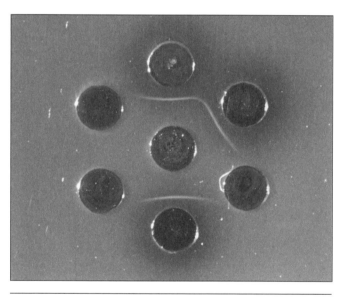

Plate 27 Gram-positive cocci growing on mannitol-salt agar plates. *Top: Micrococcus. Bottom left: Staphylococcus aureus. Bottom right: Staphylococcus epidermidis.* Courtesy of the University of Washington.

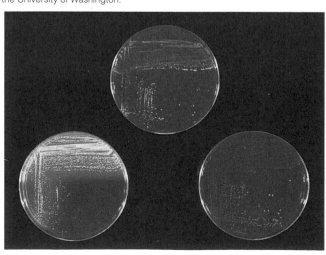

Plate 28 *Deinococcus,* Gram-positive coccus resistant to radiation. Courtesy of the University of Washington.

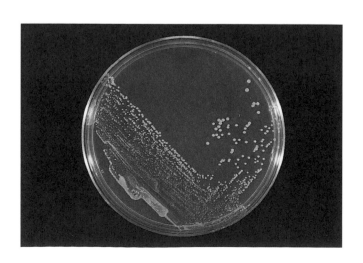

*I*ntroduction to Microbial Genetics

In this section, three aspects of microbial genetics will be studied: selection of mutants, gene transfer, and gene regulation.

Selection of Mutants Mutations are constantly occurring in all living things. The replication of DNA is amazingly error-free but about once in every 100 million duplications of a gene a change is made. There are three possible outcomes:

1. There will be no effect. Perhaps the altered base did not lead to structural change in a protein and the cell remained functional.
2. The mutation may have affected a critical portion of an essential protein resulting in the death of the cell.
3. In rare cases the mutation may enable the cell to grow faster or survive longer than the other nonmutated cells.

Gene Transfer Bacteria can transfer genetic material to other bacteria in three ways: conjugation, transduction, and transformation. Conjugation occurs during cell-to-cell contact and is somewhat similar to sexual recombination seen in other organisms. The transferred DNA can be either chromosomal or a small, circular piece of DNA called a plasmid.

Transduction is the transfer of genes from one bacterial cell to another by a bacterial virus. These viruses, called bacteriophages or phages, package a bacterial gene along with the viral genes and transfer it to a new cell.

A third method of transferring genes is transformation, which is also called DNA-mediated transformation. (The word *transformation* is sometimes used to define the change of normal animal cells to malignant cells—a completely different system.) In bacterial transformation, isolated DNA is mixed with viable cells. It then enters the cells, which are able to express these new genes. Although it would seem to be impossible for a large molecule such as DNA to enter through the cell wall and membrane of a living cell, this is indeed what happens.

Gene Regulation Another aspect of genetics is the expression of genes. A cell must be economical with its energy and material, and must not make enzymes or other products when they are not needed. On the other hand, a cell must be able to "turn on" genes when they are required in a particular environment. Gene regulation is examined in exercise 18.

NOTES:

EXERCISE

Selection of Bacterial Mutants Resistant to Antibiotics

Getting Started

All the bacterial cells in a pure culture are derived from a single cell. These cells, however, are not identical because all genes tend to mutate and form mutant organisms. The spontaneous **mutation rate** of genes varies between 1 in 10^4 to 1 in 10^{12} divisions, and even though that is quite a rare event, significant mutations are observed because bacterial populations are very large. In a bacterial suspension of 10^9 cells/ml, one could expect 10 mutations of a gene that mutated 1 in every 10^8 divisions.

Mutant bacteria usually do not grow as well as the **wild-type** normal cell because most changes are harmful, or at least not helpful. If, however, conditions change in the environment and favor a mutant cell, it will be able to outcompete and outgrow the cells that do not have the advantageous mutation. It is important to understand that the mutation is a random event that the cell cannot direct. No matter how useful a **mutation** might be in a certain situation, it just happens to the cell, randomly conferring an advantage or disadvantage to it.

In this exercise, you select bacteria resistant to streptomycin. Streptomycin is an **antibiotic** that kills bacteria by acting on their ribosomes to prevent protein synthesis. (However, it does not stop protein synthesis in animals because eukaryotic ribosomes are larger than those of bacteria and therefore different.) **Sensitive** E. coli cells can become resistant to streptomycin with just one mutation.

In this exercise, you select organisms resistant to streptomycin by adding a large population of **sensitive** bacteria to a bottle of TS broth containing streptomycin. Only organisms that already had a random mutation for streptomycin resistance will be able to survive and multiply (figure 15.1).

Figure 15.1 Selection of streptomycin-resistant E. coli cells.

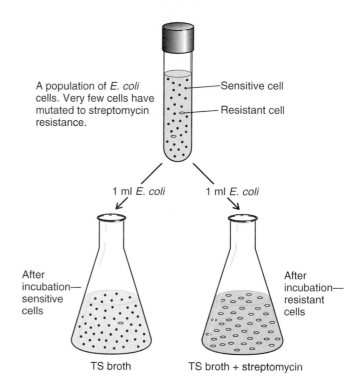

A population of E. coli cells. Very few cells have mutated to streptomycin resistance.

Sensitive cell

Resistant cell

1 ml E. coli

1 ml E. coli

After incubation—sensitive cells

After incubation—resistant cells

TS broth

TS broth + streptomycin

Definitions

Antibiotic. A substance produced by one organism, usually a microorganism, which kills or inhibits other organisms.

Mutation. An inheritable change in the base sequences of DNA.

Mutation rate. The number of mutations per cell division.

Sensitive. An organism killed or inhibited by a particular antibiotic.

Wild type. The organism as it is isolated from nature.

Objectives

1. To understand the concept of selection and its relationship to mutation.
2. To understand that mutations are random events, and that the cell cannot cause specific mutations to occur, no matter how advantageous they might be.
3. To count the number of streptomycin-resistant mutant bacteria that occur in an overnight culture of a sensitive strain.

Reference

Nester et al. *Microbiology: A human perspective,* 4th ed., 2004. Chapter 8, Section 8.6.

Materials

Per team

First Session

Flasks (or bottles) containing 50 ml TS broth, 2

TS agar deeps, 2

Sterile petri dishes, 2

1-ml pipets, 2

Overnight broth culture (~ 18 hours) of *Escherichia coli* K12 (about 10^9 cells/ml)

Streptomycin solution at 30 mg/ml

Second Session

TS agar deeps, 2

Sterile petri dishes, 2

1-ml pipets, 2

Tubes of 0.5 ml sterile water, 2

0.1-ml streptomycin

Procedure

First Session

1. Melt and place 2 TS agar deeps in a 50°C water bath.
2. Label one petri plate and one flask "with streptomycin." Label the other flask and plate "without streptomycin control" (figure 15.2).
3. Add 0.3 ml streptomycin to the flask labeled streptomycin and 0.1 ml to one of the melted cooled agar deeps. Discard the pipet.
4. Immediately inoculate the agar deep with 1 ml of the bacterial culture, mix, and pour in the plate labeled "with streptomycin."
5. Add 1 ml of the bacteria to the tube of melted, cooled agar without streptomycin and pour into plate labeled "without streptomycin."
6. Add 1 ml of bacteria to each of the flasks.
7. Incubate the plates and flasks at 37°C. If using bottles lay them on their side to increase aeration.

Second Session

1. Melt and cool two tubes of TS agar in 50°C water bath.
2. Pour one tube of melted agar into a petri dish labeled "without streptomycin" and let harden.

Figure 15.2 Inoculating media with and without streptomycin with a culture of *E. coli* (Session 1).

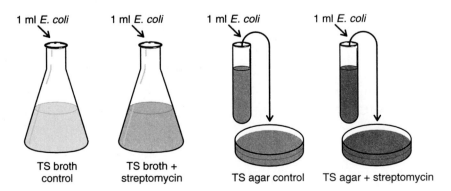

| 1 ml *E. coli* | 1 ml *E. coli* | 1 ml *E. coli* | 1 ml *E. coli* |

TS broth control   TS broth + streptomycin   TS agar control   TS agar + streptomycin

3. Add 0.1 ml streptomycin to the other tube of melted agar, pour into a petri dish labeled "with streptomycin" and let harden.
4. Examine the bottles and plates inoculated last period. Note whether there is growth (turbidity) or not in both of the bottles. Count the number of colonies growing in the pour plates. How many streptomycin-resistant mutants/ml were present in the original inoculum? Compare it to the growth of organisms in the control plate without streptomycin. If there are more than 300 colonies or the plate is covered by confluent growth, record as TNTC or "too numerous to count." Record results.
5. Test the bacteria growing in the bottles and on the plates for sensitivity or resistance to streptomycin in the following way. Divide both agar plates in four sections as diagrammed in figure 15.3. Take a loopful of broth from the bottle without streptomycin and inoculate a

sector of each agar plate. Do the same with the broth culture containing streptomycin.
6. Dig an isolated colony out of the agar plate containing the streptomycin and suspend it in a tube of sterile water. Use a loopful to inoculate the third sector of each plate. Also suspend some organisms from the control plate in saline (there will not be any isolated colonies) and inoculate the fourth sector. Incubate the plates at 37°C.
7. Predict which bacteria will be sensitive to streptomycin and which will be resistant.

Third Session

1. Observe growth on each sector of the plates and record results. Were they as you predicted?
2. Occasionally, mutants will not only be resistant to streptomycin, but also will require it. If you have one of these unusual mutants, be sure to show it to the instructor.

Figure 15.3 Testing above incubated cultures for streptomycin sensitivity (Session 2).

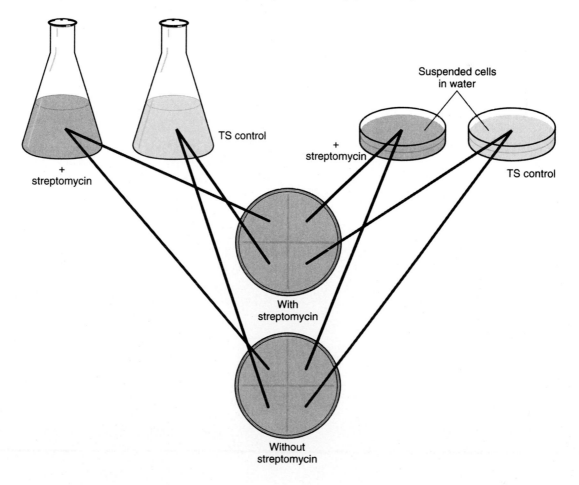

NOTES:

EXERCISE

Laboratory Report:
Selection of Bacterial Mutants Resistant to Antibiotics

Results (After Second Session)

Source	Growth / No Growth
TS broth (control)	
TS broth plus streptomycin	

Source	Number of colonies
TS agar plate (control)	
TS agar plate plus streptomycin	

Results (After Third Session)

Source	Growth on TS Agar Plate	Growth on TS Agar Plate + Strp
TS broth (control)		
TS broth plus streptomycin		
TS agar (control)		
TS agar plus streptomycin		

How many organisms/ml were streptomycin resistant in the original overnight culture of sensitive *E. coli?*

Questions

1. Two bottles of TS broth (with and without streptomycin) were inoculated in Session 1 with 1 ml of an overnight culture of *E. coli*. After incubation, why was one population streptomycin sensitive and the other streptomycin resistant?

2. How were you able to estimate the number of streptomycin-resistant organisms already present in the overnight culture of *E. coli* growing in the TS broth?

3. Why should antibiotics not be used unless they are necessary?

4. Which is correct?
   a. An organism becomes resistant after it is exposed to an antibiotic.

   b. An antibiotic selects organisms that are already resistant.

EXERCISE

Transformation: A Form of Genetic Recombination

Getting Started

In this exercise, transformation is used to transfer the genes of one bacterium to another. It gives you a chance to see the results of what seems to be an impossible process—a huge DNA molecule entering an intact cell and permanently changing its genetic makeup.

Basically, the process involves mixing DNA from one strain of lysed (disrupted) cells with another strain of living cells. The DNA then enters the viable cells and is incorporated into the bacterial chromosome. The new DNA is expressed and the genetic capability of the cell may be changed.

In order to determine whether the bacteria are indeed taking up additional DNA, the two sets of organisms (DNA donors and DNA recipients) must differ in some way. One strain usually has a "marker" such as resistance to an antibiotic, or the inability to synthesize an amino acid or vitamin. In this exercise, a gene responsible for conferring resistance to the antibiotic streptomycin is transferred to cells that are sensitive to it (figure 16.1).

The organism used in this exercise is *Acinetobacter* (ā sin NEET o bacter), a short, Gram-negative rod found in soil and water. The prefix "a" means without, and "cine" means movement, as in cinema; thus *Acinetobacter* is nonmotile. This organism is always **competent** which means it can always take up **naked DNA**. Other organisms are not competent unless they are in a particular part of the growth curve or in a special physiological condition. The DNA must not be degraded for transformation to take place in any event. If an enzyme such as **DNase** is present, it cuts the DNA in small pieces preventing transformation.

Definitions

Competent. Cells that are able to take up naked DNA.

Figure 16.1 Transformation of cells with a gene conferring streptomycin resistance.

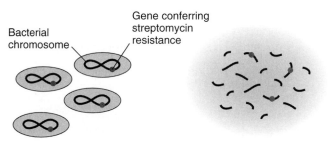

Cells resistant to streptomycin   Cells lysed, releasing DNA

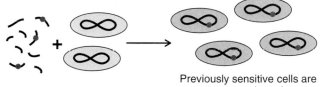

DNA mixed with sensitive cells

Previously sensitive cells are now streptomycin resistant. They have been transformed.

DNase. An enzyme that cuts DNA, making it useless for transformation.

Naked DNA. DNA released from lysed or disrupted cells and no longer protected by an intact cell.

Objectives

1. To understand the process of transformation and observe it in the laboratory.
2. To understand the use of genetic markers.
3. To understand the importance of controls in an experiment.

Reference

Nester et al. *Microbiology: A human perspective,* 4th ed., 2004. Chapter 8, Section 8.7.

Safety Precaution: *Acinetobacter* can cause pneumonia in immunologically compromised individuals.

Materials

Per team

    TSY agar plate, 1

    TSY agar plate with streptomycin (second session), 1

    Broth culture of *Acinetobacter* Str^R (resistant to streptomycin), 1

    Broth culture of *Acinetobacter* Str^S (sensitive to streptomycin), 1

    Tube with 0.1 ml detergent SDS (sodium dodecyl sulfate) in 10× saline citrate, 1

    Solution of DNase

    1-ml pipet, 1

Class equipment

    60°C water bath with test tube rack

Procedure

First Session

1. Transfer 1.0 ml of Str^R *Acinetobacter* broth culture into the tube of SDS. Label the tube and incubate it in the 60°C water bath for thirty minutes. The detergent (SDS) will lyse the cells, releasing DNA and other cell contents. Any cells that are not lysed will be killed by the thirty-minute exposure to 60°C water. Label the tube DNA.

2. Divide the bottom of the TSY agar plate into 5 sectors using a marking pen. Label the sections DNA, Str^S (streptomycin sensitive), Str^R (streptomycin resistant), Str^S + DNA, and Str^S + DNase + DNA (figure 16.2).

3. Inoculate the plate as indicated by adding a loop of the broth culture, DNA, or DNase in an area about the size of a dime to each sector.

   a. DNA. The lysed mixture of Str^R cells is the source of DNA. It also contains RNA, proteins, and all the other cell components of the lysed cells, which do not interfere with the transformation. This mixture is inoculated onto the TSY agar plate to show that it contains no viable organisms. It is a control.

Figure 16.2 (*a-e*) Five labeled sectors of a TSY agar plate.

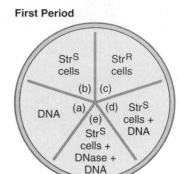

TSY agar

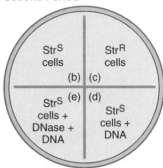

TSY agar + streptomycin

   b. Str^S cells. Inoculate a loopful of the Str^S culture. This step should demonstrate that the Str^S cells can grow on TSY agar (which is a control).

   c. Str^R. Inoculate a loopful of the Str^R culture. This step should demonstrate that the Str^R culture is viable (it is a control).

   d. Str^S cells + DNA. Inoculate a loopful of Str^S cells and add a loopful of the DNA (lysed Str^R cells) in the same area. THIS IS THE ACTUAL TRANSFORMATION. Str^R cells will grow here if transformed by the DNA.

   e. Inoculate a loopful of Str^S cells as above, and in the same area add a loopful of DNase, then add a loopful of DNA. It is important to add these in the correct order (otherwise transformation will occur before the DNase can be added). This is a control to demonstrate that DNA is indeed the component of the lysed cells

      16–2 Exercise 16 Transformation: A Form of Genetic Recombination

that is responsible for the transformation. The DNase should inactivate the DNA, preventing transformation.

4. Incubate the plates at room temperature for several days or at 37°C for 48 hours.

Second Session

1. Observe the plate you prepared in the first session. There should be growth in all sectors of the plate except the DNA sector (*a*). If the DNA control sector shows growth, it indicates that your crude DNA preparation was not sterile but contained viable cells. If this has happened, discard your plates and borrow another student's plate after he or she is finished with it; there should be sufficient material for more than one team. Why is it so important that the DNA preparation is sterile?

2. Divide the bottom of a TSY + streptomycin plate into four sectors and label them Str^S, Str^R, Str^S + DNA, and Str^S + DNase + DNA.

3. Streak a loopful of cells from the first plate to the corresponding sectors on the TSY + streptomycin plate. Lightly spread them in an area about the size of a dime. Cells growing on this plate must be streptomycin resistant.

4. Incubate at room temperature for several days or at 37°C for 48 hours or until cells have grown.

Third Session

1. Observe the TSY + streptomycin agar plate inoculated last period and record results. Did you transform the cells sensitive to streptomycin to cells that were resistant and could now grow on streptomycin?

NOTES:

EXERCISE

Laboratory Report: Transformation:
A Form of Genetic Recombination

Results

Indicate growth (+) or no growth (−) in each sector.

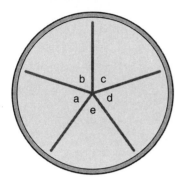

TSY agar

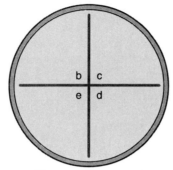

TSY agar + streptomycin

	Yes/No	Sector that Demonstrates Observation
Were *Acinetobacter* Str^S cells sensitive to streptomycin?		
Were *Acinetobacter* Str^R cells resistant to streptomycin?		
Was the DNA (cell lysate) free of viable cells?		
Did transformation take place?		
Did the DNase prevent transformation?		

Questions

1. What two components were mixed together to show transformation?

2. What is the action of DNase?

3. What control showed that transformation and not conjugation or transduction was responsible for the results?

4. If the Str^S cells had grown on the TSY + streptomycin agar, would you have been able to determine if transformation had taken place? Explain.

5. If you had used a DNA lysate containing viable cells, would it have been possible to determine whether transformation had taken place? Explain.

6. How does transformation differ from conjugation and transduction?

EXERCISE 17

17

Bacterial Conjugation

Getting Started

Conjugation is one of the three mechanisms responsible for genetic transfer in bacteria (see Introduction). In conjugation, a cell of mating type F$^+$ attaches by a pilus to a cell of mating type F$^-$ and DNA is transferred from the F$^+$ cell to the F$^-$.

The donor cell is designated F$^+$ because it contains a fertility factor known as an **F factor** and the recipient is F$^-$ because it lacks the F factor. Some of the genes that make up the F factor code for the sex pilus, an appendage on the surface of the cell.

The F factor is usually located on a **plasmid,** which is a small, circular piece of DNA found in the cytoplasm of the cell. The genes found on plasmids are usually helpful to the cell in certain situations, but are not essential for the cell's normal metabolism. For example, the genes conferring a certain type of resistance to antibiotics are frequently carried on plasmids.

During conjugation, the pilus of the F$^+$ cell attaches to the F$^-$ cell and seems to be involved in bringing the cells into close contact. One strand of the plasmid enters the F$^-$ cell, the plasmid DNA replicates, and the F$^-$ cell becomes an F$^+$ cell. The cell can then express all the other genes (such as antibiotic resistance) contained on the plasmid. Conjugation is therefore a very important mechanism in the spread of antibiotic-resistant genes (figure 17.1*a*).

F factors can be responsible for the transfer of more than just the genes on a plasmid. The F factor can, on rare occasions, integrate into the bacterial chromosome, which can then be transferred during conjugation (see figure 17.1*b*). These strains are called **Hfr** strains for *high frequency of recombination.*

In this exercise, a donor Hfr strain containing the F factor in the chromosome is used to transfer chromosomal genes to an F$^-$ strain of *E. coli.* Each of the strains must have a genetic "marker" such as antibiotic resistance or inability to synthesize an es-

sential cellular metabolite such as an amino acid or vitamin. In this case, the donor strain is unable to synthesize the amino acid methionine and the recipient cannot synthesize the amino acid threonine. Neither strain can grow on a mineral salts medium because it does not contain the amino acids that the strains require. (Mutants requiring a growth factor are termed **auxotrophs.**) If, however, the normal genes are transferred to the mutant recipient during conjugation, these cells then will be able to synthesize all the required amino acids. Recombinant bacteria will be able to grow on the mineral salts medium but the auxotrophic parent strains will not.

During conjugation, the chromosome is transferred in a linear manner with the F factor at the end—similar to a caboose. The longer the cells stay in contact, the more chromosome is transferred, but the cells usually break apart before the whole chromosome is transferred. It is important not to shake the culture during conjugation so that cells stay in contact as long as possible.

Definitions

Auxotroph.   An organism that cannot synthesize all its needed growth factors. The strain labeled meth- in the conjugation exercise is an auxotroph requiring methionine. It cannot grow on a mineral medium, but only on TS agar or another medium that contains methionine.

Conjugation.   A method of transferring DNA between bacteria requiring cell-to-cell contact.

F factor.   (F plasmid) Genes giving the cell the ability to transfer DNA via conjugation.

F-.   Cells lacking the F factor, and are called the recipient bacteria.

F+.   Cells containing the F factor, and are called donor bacteria.

Figure 17.1 *(a)* Conjugation: transfer of the F⁺ plasmid. *(b)* Conjugation: transfer of chromosomal DNA.

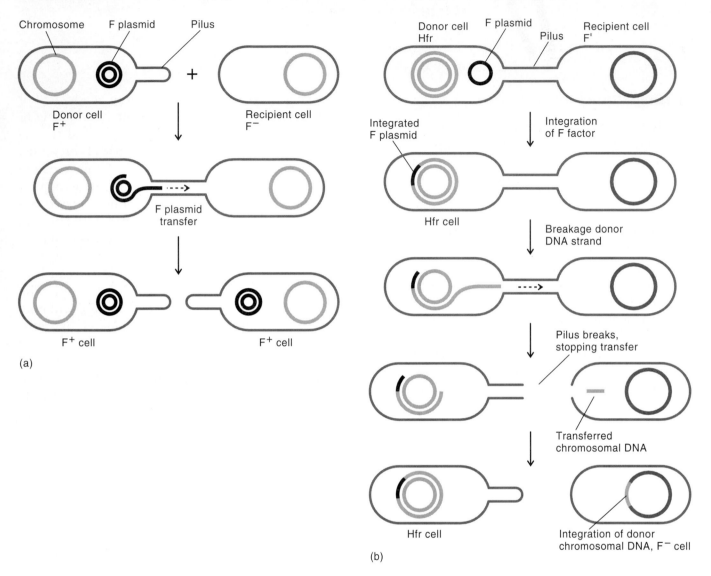

Hfr. (High frequency of recombination) F factor is incorporated in the chromosome of the bacteria, and consequently portions of the chromosome can be transferred to F⁻ bacteria.

Plasmid. A small circle of DNA found in some cells in addition to the chromosomal DNA.

Objectives

1. To increase knowledge of concepts and techniques used in the study of genetics.

2. To acquaint you with a laboratory method for demonstrating conjugation using auxotrophic organisms.
3. To demonstrate the importance of conjugation in transferring genes among bacteria.

Reference

Nester et al. **Microbiology: A human perspective,** 4th ed., 2004. Chapter 8, Section 8.9.

Materials

Per team

TS broth culture in log phase of *E. coli* A F+ methionine-(Hfr donor)

TS broth culture in log phase of *E. coli* B F− threonine-(recipient)

TS agar plates, 1

Sterile test tube, 1

Sterile 9-ml water blanks, 5

Agar plates of mineral salts + 0.5% glucose (mineral medium or MM), 4

Sterile 1.0-ml pipets, 6

Sterile bent glass rod (sometimes called dally rods or hockey sticks), 1

Procedure

Steps are outlined in figure 17.2.

First Session

1. Label all tubes and plates as indicated in figure 17.2.
2. Transfer 1 ml of culture A into the water blank labeled A 10^{-1}. With the same pipet, transfer a second 1 ml of culture A into the tube labeled A + B.
3. Transfer 1 ml of culture B into the water blank labeled B 10^{-1}. With the same pipet, transfer a second 1 ml of culture B into the tube labeled A + B.
4. Gently mix the A + B tube and incubate it without shaking for 30 minutes at room temperature. This is the actual mating.
5. While the A + B mixture is incubating, inoculate half of the MM plate (labeled A) and half of the TS agar plate (labeled A) with loopfuls of culture A. Repeat with culture B. These are controls to verify that neither culture A nor B can grow on MM agar, but can grow on TS agar. These cultures were diluted 1/10 so that you would not carry over any of the rich TS broth medium when testing the organisms on the minimal medium.
6. After 30 minutes incubation, pipette 1 ml of the A + B mixture into the water blank labeled A + B 10^{-1}. Mix thoroughly and transfer 1 ml with a second sterile pipet to the water blank labeled A + B 10^{-2}. Again, mix thoroughly and transfer 1 ml with a third sterile pipet to the water blank labeled A + B 10^{-3}. Mix thoroughly.
7. With a sterile 1.0-ml pipet remove 0.1 ml from the A + B 10^{-3} dilution and add it to the surface of the MM plate labeled A + B 10^{-3}. Immediately spread the drop completely over the surface of the plate with the sterile bent glass rod.
8. Using the same pipet and bent glass rod, repeat the procedure to inoculate the MM agar plates labeled A + B 10^{-2} and A + B 10^{-1} from their respective dilutions. Remember, you can use the same pipet when going from low concentrations to higher concentrations.
9. Invert the agar plates and incubate at 37°C for 2–3 days.

Second Session

1. Examine the control MM and TS agar plates and record results. Are both cultures A and B viable auxotrophs?
2. Count the number of colonies on the plates that have countable numbers. Record the results.

Figure 17.2 Procedure for conjugation.

Conjugation

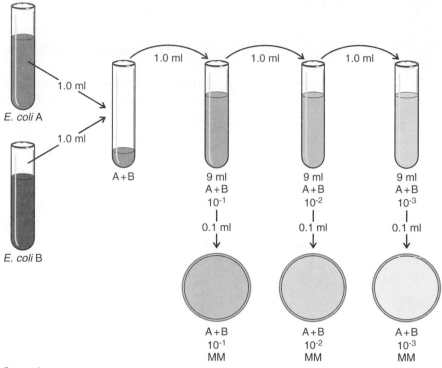

Controls

Use the same tubes of *E. coli* A and *E. coli* B shown above.

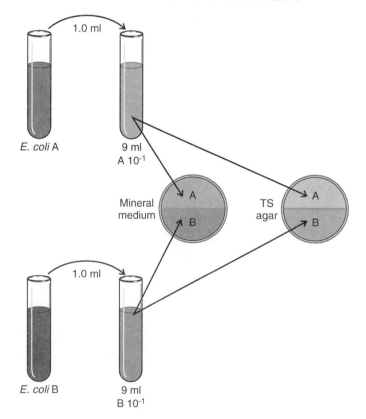

EXERCISE

Laboratory Report: Bacterial Conjugation

Results

Indicate where growth occurred.

	Culture A		Culture B	
TS agar				
MM agar				
Culture an auxotroph?				

Plate Counts	A + B 10^{-1}	A + B 10^{-2}	A + B 10^{-3}
Number of colonies			

1. How many recombinant organisms resulted from the mating? Show calculations.

2. If there were roughly 2×10^8 organisms in the A + B mixture, what percent of the original mixture resulted in recombinants?

   Note: Since only the recipient cells can receive DNA and change genotype, the percent of recombinants should be based on the number of recipient cells, or 1×10^8 cells/ml.

3. If the back mutation rate for threonine is about 1 in 10^8 cells, about how many back mutants would you expect to be present per ml in the A + B mixture?

Questions

1. Why can the same pipet and bent glass rod be used to inoculate plates when starting with the most dilute mixture?

2. Why is an auxotrophic organism not able to grow on MM agar?

3. If either A or B could grow on the MM agar, how would that change the results?

4. Using the answers you calculated in part 2 in the Results section, how did the number of recombinants resulting from conjugation compare with the number you expect from back mutation?

5. Compare the effects of mutation and conjugation to produce organisms with new genetic capability.

NOTES:

EXERCISE

Gene Regulation: Induction and Catabolite Repression

Getting Started

A bacterial cell has all the genetic information to produce and operate a new cell. This includes the enzymes necessary for obtaining energy and synthesizing the necessary cellular components. A cell must work as efficiently as possible and carefully utilize the available nutrients without investing energy in enzymes not needed. Some enzymes are utilized in the basic energy pathways of the cell and are continually synthesized. These are called **constitutive enzymes.**

Others termed **inducible enzymes** are only needed when their specific **substrate** is available. For example, *E. coli* can break down the sugar lactose with an enzyme called β-galactosidase. If lactose were not present in the environment, it would be a waste of energy and of intermediate compounds to synthesize this enzyme. Therefore, these kinds of enzymes are called inducible enzymes because the presence of the substrate induces their synthesis.

How does the cell control inducible enzymes? In bacteria, this takes place on the level of **transcription.** Inducible enzymes can be found in an **operon,** which has a promoter and operator followed by the genes involved in the enzymatic activity (figure 18.1). The repressor binds to the operator gene blocking transcription if the substrate for the enzyme is not present.

When the substrate is present, it binds to the repressor allowing the RNA polymerase to transcribe the genes for β-galactosidase.

Another method of control utilized by the cell is catabolite repression, which occurs when a cell has a choice of two sources of energy, one of which is more easily utilized than the other. An example is the presence of both glucose and starch. Glucose can immediately enter the glycolytic pathway, while starch must be first cleaved with amylase. Amylases and other enzymes cost the cell energy and materials to produce, so therefore it is much more economical for the cell to utilize glucose if it is present. When glucose is present along with starch, the glucose represses the synthesis of amylase even though the enzyme would normally be induced in the presence of the starch.

The first procedure investigates induction of β-galactosidase, a particularly important enzyme system used in recombinant DNA techniques as a measure of gene expression.

The second procedure tests for the catabolite repression of amylase. Amylase is an exoenzyme that is excreted outside the cell because the large starch molecule may be too large to pass through the cell membranes.

Definitions

Constitutive enzymes. Enzymes continually produced by the cell.

Inducible enzymes. Enzymes produced only when substrate is present.

Operon. A series of genes that is controlled by one operator (gene).

Substrate. The molecule reacting with an enzyme.

Transcription. Transfer of the genetic information from DNA to messenger RNA.

Objectives

1. To understand the concepts of induction.
2. To understand the lac operon and the use of ONPG.
3. To understand the concept of catabolite repression and how it is tested.

Reference

Nester et al. *Microbiology: A human perspective,* 4th ed., 2004. Chapter 7, Section 7.6.

Figure 18.1 Inducible enzymes. (*a-b*) Regulation of enzyme synthesis in a degradative enzyme system. Note that the same genetic elements take part as in the regulation of a biosynthetic pathway. From Eugene W. Nester et al. *Microbiology: A Human Perspective.* Copyright © 1998 The McGraw-Hill Companies. All Rights Reserved. Reprinted by permission.

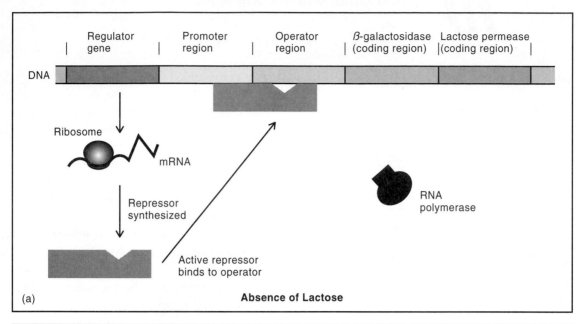

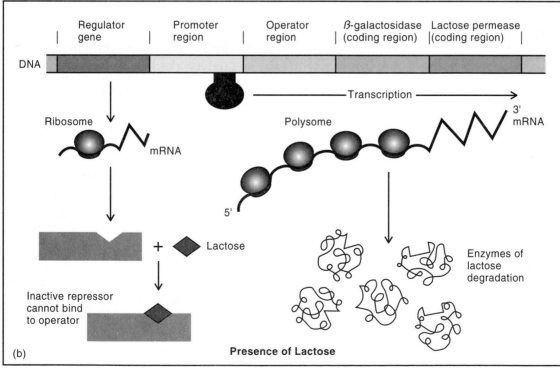

Induction

Procedure

First Session

1. Inoculate each tube with a drop or loopful of *E. coli*.
2. Incubate at 37°C until the next laboratory period or at least 48 hours.

Second Session

1. Examine the tubes for growth—all the tubes should be turbid.
2. Add 1 ml of ONPG to each tube. The indicator compound ONPG is cleaved by β-galactosidase into a yellow product.
3. Incubate at room temperature for 30 minutes.
4. Examine the tubes to determine if the broth has turned yellow—an indication of the presence of the induced enzyme β-galactosidase.
5. Record the results.

Catabolite Repression

Procedure

First Session

1. Label each plate.
2. Inoculate the middle of each plate with the *Bacillus* in an area about a few mm square.
3. Incubate at 30°C for a day or two. Try not to let the *Bacillus* grow over more than a third or half the plate. You may have to refrigerate the plates if the colony becomes too large.

Second Session

1. Flood the agar plate with Gram's iodine. The starch will turn purple. If the starch has been broken down with amylases, a clear zone will appear around the colony (figure 18.2).
2. Record the results.

Figure 18.2 Appearance of plates after flooding with Gram's iodine.

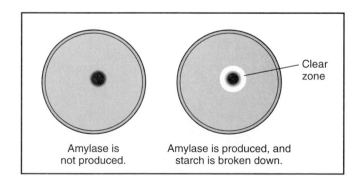

Amylase is not produced.   Amylase is produced, and starch is broken down.

Clear zone

NOTES:

EXERCISE

Laboratory Report: Gene Regulation: Induction and Catabolite Repression

Results

Results of induction

	Glucose	Lactose	Glycerol
Color			
β-galactosidase present?			

Results of catabolite repression

	Starch	Starch + Glucose
Zone of clearing present		
Presence of amylase		

Questions

1. Which substrate induced β-galactosidase?

2. What reaction produced the yellow color?

3. What results would you expect if β-galactosidase were a constitutive enzyme?

4. In the catabolite repression exercise, did the *Bacillus* have the capacity to synthesize amylase (amylase +)? How did you determine?

5. Did you observe catabolite repression when glucose was added to the starch? How did you determine?

6. What results would you expect if amylase were a constitutive enzyme?

Introduction to the Other Microbial World

The phrase, the "other microbial world" refers to organisms other than bacteria, the major organisms of study in other parts of the manual. All of the organisms included here, with the exception of viruses, are eukaryotic organisms, many of which are of medical importance. Included are members of the nonfilamentous fungi (primarily yeasts), the filamentous fungi (molds), and intestinal animal parasites of medical importance, the protozoa and helminths. Viruses that infect both prokaryotes (bacteria) and eukaryotes (animal and plant cells) are introduced. Only bacterial viruses will be made available for laboratory study.

Mycology, or the study of fungi, is the subject of exercise 19. Included are a wide variety of forms, ranging from unicellular microscopic cells, such as yeast, to filamentous multicellular structures. Many filamentous fungi such as mushrooms, puffballs, toadstools, bracket fungi, and molds are visible with the naked eye.

Most yeasts and filamentous fungi found in nature are nonpathogenic. In fact, many contribute to our well-being; for example, the yeast *Saccharomyces* is important for manufacture of bread, beer, and wine; and the filamentous fungus *Penicillium chrysogenum* produces the antibiotic penicillin. Some are associated with spoilage (moldy jam and bread, mildew of clothing, etching of glass lenses in warm, humid environments) and, more recently, with production of food toxins known as aflatoxins by certain species of the genus *Aspergillus.*

Fungi are eukaryotic organisms which differ from algae in that they lack chlorophyll. Fungi differ from bacteria in that the cells are larger and contain membrane-bound organelles. In addition, bacteria are prokaryotes.

The latter difference, in turn, affects antibiotic therapy; for example, antibiotics effective against prokaryotes are often ineffective against eukaryotes. If an antibiotic is effective against a fungus it may also, depending on the mode of action of the antibiotic, damage the human host, because humans too are eukaryotic. Consequently, antibiotic control of fungal infections is usually more difficult than control of bacterial infections. Fortunately, many fungal infections are opportunistic[1] infections so that healthy individuals rarely acquire them other than, perhaps, cutaneous fungal infections such as athlete's foot. Nevertheless, there are a number of important mycotic diseases with which you should familiarize yourself.

Fungi can be cultivated in the laboratory in the same manner as bacteria. Physiologically, all fungi are heterotrophs (they require an organic source of carbon, such as glucose) and most are aerobic, although some are facultative, that is, able to carry out both aerobic and anaerobic metabolism. Most fungi grow best at temperatures of 20° to 30°C, although some grow well at temperatures as high as 45° to 50°C (such as *Aspergillus fumigatus,* an opportunistic filamentous fungus known to cause pulmonary aspergillosis).

Parasitic diseases constitute a major worldwide public health problem, both in developed and developing countries. In developing countries, parasitic diseases are prominent due to poverty, malnutrition, lack of sanitation, and lack of education. A simple family survey for intestinal parasites in a small Central American village revealed that every family member harbored at least three types of parasites. Effects of human parasitic disease range from minimal, with some nutritional loss but only minor discomfort (common in pinworm and *Ascaris* infections), to debilitating and life-threatening disorders such as malaria and schistosomiasis.

Fecal contamination of drinking water by wild animal carriers such as beavers has in recent years caused major outbreaks of **giardiasis** (an intestinal disease) in various parts of the United States. The causative agent, the protozoan *Giardia lamblia* (see figure 33.3), produces **cysts** that are quite small,

---

[1]Opportunistic infections are associated with debilitating diseases (such as cancer) and use of cytotoxic drugs, broad-spectrum antibiotics, and radiation therapy, all of which can suppress the normal immune response.

thereby enabling passage sometimes through faulty water supply filters. The cysts often resist chlorination. The protozoan flagellate, *Trichomonas vaginalis*, is a common cause of vaginitis in women and is often sexually transmitted. Pinworm infestations are a problem among elementary school children. About 20% of infections in domestic animals are caused by protozoan and helminthic (worm) agents.

Because of their worldwide public health importance and their natural history differences from bacteria and fungi, we believe protozoa and helminths merit their own laboratory session (see exercise 20). It provides insights on the diseases they cause and the techniques used for their diagnosis and identification.

Virology, the study of viruses (the word for poison in Greek), also had early roots, although somewhat mysterious. Mysterious, in part, because the viruses could not be seen, even with a light microscope, and yet when sap from the leaf of an infected tobacco plant was passed through a filter that retained bacteria and fungi, the clear filtrate retained its infectious properties. It was not until the mid-1930s that viruses were first observed with the advent and aid of the electron microscope. In 1935, Wendell Stanley succeeded in crystallizing tobacco mosaic virus (TMV), enabling him to observe that it was structurally different from living cells (figure I.7.1).

Bacterial viruses known as **bacteriophage** were first described by Twort (1915) and later by d'Herelle (1917). d'Herelle observed their filterable nature and their ability to form **plaques** on an agar plate seeded with a lawn of the host bacterium (see figure 21.2). Both Twort and d'Herelle worked with coliform bacteria isolated from the intestinal tract.

The discovery of plaque formation on an agar medium was a godsend for future virology research in that it provided a fast, easy way to recognize, identify, and quantify bacterial virus infections. Viruses that attack mammalian cells also form structures analogous to plaques when cultivated on growth media able to support mammalian cell growth. Rather than plaques, the structures formed are described as **cytopathic effects (CPE)** (figure I.7.2). The CPE observed is dependent upon the

Figure I.7.1 Tobacco mosaic virus. Electron micrograph (X approximately 70,000). Compare length and width with a rod-shaped bacterium. © Omikron/Photo Researchers, Inc.

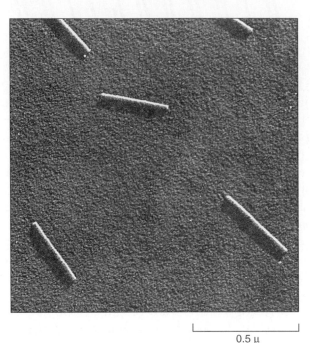

0.5 μ

Figure I.7.2 Mammalian virus plaques showing different cytopathic effects. Photographed 5 days after infection of the growth medium (a single layer of monkey kidney cells) with the various mammalian viruses. When in monolayers, the viruses are able to form plaques (a form of CPE), which can be detected macroscopically. From *Diagnostic Procedures for Viral. Rickettsial and Chlamydial Infections*, 5th Edition. Copyright © 1979 by the American Public Health Association. Reprinted with permission.

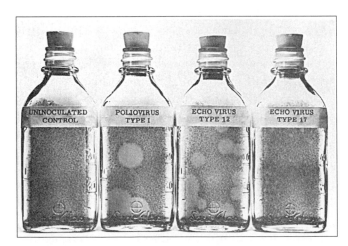

nature of both the host cell and invading virus. The same can be said about the nature of bacterial plaque formation.

The advantage of using a bacteriophage to demonstrate virus growth is the ease of culturing the host cells. Mammalian cells require complex growth media including blood serum as well as prolonged incubation with the virus before being able to observe CPE. Therefore, we will work with an *E. coli* bacteriophage that can either be isolated and concentrated from sewage or obtained from a pure culture collection (see exercise 21).

Having made sundry efforts, from time to time, to discover, if 'twere possible, the cause of the hotness or power whereby pepper affects the tongue (more especially because we find that even though pepper hath lain a whole year in vinegar, it yet retaineth its pungency); I did now place anew about ⅓ ounce of whole pepper in water, and set it in my closet, with no other design than to soften the pepper, that I could the better study it. This pepper having lain about three weeks in the water, and on two several occasions snow-water having been added thereto, because the water had evaporated away; by chance observing this water on the 24th of April,

1676, I saw therein, with great wonder, incredibly many very little animalcules, of divers sorts; and among others, some that were 3 or 4 times as long as broad, though their whole thickness was not, in my judgment, much thicker than one of the hairs wherewith the body of a louse is beset.... The second sort of animalcules consisted of a perfect oval. They had no less nimble a motion than the animalcules first described, but they were in much greater numbers. And there was also a third sort, which exceeded both the former sorts in number. These were little animals with tails, like those that I've said were in rainwater.

The fourth sort of little animals, which drifted among the three sorts aforesaid, were incredibly small; nay, so small, in my sight, that I judged that even if 100 of these very wee animals lay stretched out one against another, they could not reach to the length of a grain of coarse sand; and if this be true, then ten thousand of these living creatures could scarce equal the bulk of a coarse sand-grain.

I discovered yet a fifth sort, which had about the thickness of the last-said animalcules, but which were near twice as long.

DOBELL, *Antony van Leeuwenhoek and his Little Animals*

NOTES:

EXERCISE 19

Microscopic Identification of Fungi

Getting Started

When you hear the word "fungus" what might it suggest? If unacquainted with fungus nomenclature, it may make you think of mushrooms and **toadstools,** or moldy fruit and yeast. All of these are examples of fungi. Fungi such as mushrooms and toadstools are macroscopic fungi that can usually be identified without the aid of a microscope (for examples see color plate 8), whereas yeast and molds require microscopy for identification (see color plates 6, 9, and 12 through 17). Since fungi are considerably larger than bacteria they may be easier to identify. Compare cell size of bacterial color plates 3 (×1,000) and 4 (×600) with yeast color plate 6: left (×1,000), right (×500).

Molds, yeasts, and perhaps another group, the **lichens,** are all members of the true fungi (Eumycota). The lichens are placed with the fungi for convenience because they represent dual thallus plants composed of an alga and a fungus. The two fungus subgroups are the **nonfilamentous fungi** exemplified by the yeasts which are unicellular, and the **filamentous fungi** exemplified by the molds which are multicellular and have true filaments (**hyphae**) that are either nonseptate (**coenocytic**) or septate (figure 19.1). The nonseptate filaments are multinucleate whereas the septate filaments contain either one or more nuclei per unit. This structural difference is important taxonomically in that one of the four classes of fungi, the Zygomycetes (table 19.1) is distinguished from the other classes by its lack of septate hyphae. Also in contrast to the other three classes it contains only a few human pathogens but numerous plant pathogens. Some authors divide the Zygomycetes into two classes, the Zygomycetes and the Oomycetes. The Zygomycetes are terrestrial fungi and the Oomycetes are aquatic fungi containing the preponderance of plant pathogens. Fungus classification, although still somewhat in a state of flux, continues to improve with time.

Why all this interest in fungi? As with other forms of life there are both the good and the bad fungi. Most are **saprophytes** meaning that in the chain of nature they decompose dead matter into a form which can be used to support all sorts of living matter. There are also the fungi that cause disease, in plants or animals. For example, studies show that fungi unable to synthesize certain of their own nutrients invade a plant for these nutrients and thereby destroy it. For this purpose they have spearlike hyphae that are adapted for invasive growth. Strangely enough the same type of hyphae are also involved in the formation of various multicellular organs. The multicellular organs in turn can regenerate hyphae (see Moore, 1998). Today there is mounting interest worldwide concerning the impact of fungi on plant disease. With respect

---

Figure 19.1 The two major hyphal types found in fungi.

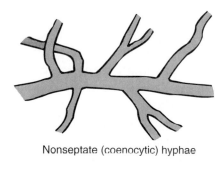

Nonseptate (coenocytic) hyphae

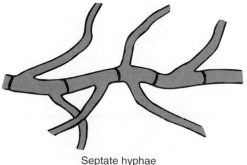

Septate hyphae

Table 19.1   Classification of the Fungi

	Class			
	Zygomycetes	Ascomycetes	Basidiomycetes	Deuteromycetes
Mycelium	Nonseptate	Septate	Septate	Septate
Sexual spores	Oospore (not in a fruiting body) found in aquatic forms; Zygospore (not in a fruiting body) found in terrestrial forms	Ascospores, borne in an ascus, usually contained in a fruiting body	Basidiospores, borne on the outside of a clublike cell (the basidium), often in a fruiting body	None
Asexual spores	Zoospores, motile; Sporangiospores, nonmotile, contained in a sporangium	Conidiospores, nonmotile, formed on the tip of a specialized filament, the conidiophore	Same as Ascomycetes	Same as Ascomycetes
Common representatives	Downy mildews, potato blight, fly fungi, bread mold (*Rhizopus*)	Yeasts, morels, cup fungi, Dutch elm disease, ergot	Mushrooms, puff balls, toadstools, rusts, smuts, stinkhorns	Mostly imperfect Ascomycetes and some imperfect Basidiomycetes*

*Some of these fungi will no doubt form sexual spores in the right environment. In this event, they would need to be reclassified.

to animals, studies show they cause disease primarily only when the animal is in a weakened condition. They are, in essence, **opportunists.**

Examples illustrating the diverse morphology of yeasts and molds when examined with the microscope are shown in figures 19.2, 19.3, 19.4, and 19.5. The beauty of fungus identification is that they may often be identified to the genus level simply by their macroscopic and microscopic growth characteristics when cultivated on various nutritional media. Most yeasts multiply vegetatively by a process known as **budding.** An exception is the genus *Schizosaccharomyces* which has the same vegetative multiplication process as bacteria—**fission** (figure 19.6). From a microscopic study of cell and bud morphology one can often determine if a yeast is a member of the genus *Saccharomyces,* or perhaps another genus with a different morphology, e.g., *Selenotila* or *Trigonopsis* (see figure 19.2). Also as earlier mentioned, fungi may be easily differentiated from bacteria by their larger size. Yeast and mold identification to the species level often requires additional morphological and physiological studies. Morphologically some yeasts form sexual spores **(ascospores)** which are borne inside an **ascus.** Examples are *Saccharomyces cerevisiae* (figure 19.7) and *Schizosaccharomyces pombe.* Most yeasts multiply asexually by budding

(see color plate 6). Some pathogenic yeasts exhibit a property known as **dimorphism.**

Examples of dimorphism are shown in table 19.2 under the column titled "Morphology." One of these, the genus *Candida,* forms oval to elongate buds when grown on the surface of Sabouraud's dextrose agar (see figure 19.2b), **chlamydospores** and **blastospores** when grown on cornmeal agar (figure 19.8), and **germ tubes** (figure 19.9) when grown in serum or egg albumin.

Examples of physiological tests used for identifying yeast to the species level are tests which evaluate their ability to **assimilate** and/or **ferment** growth media containing various sugars as the sole carbohydrate source. For assimilation studies agar plates seeded with the test yeast are inoculated on the surface with sterile paper discs containing the various sugars. Growth adjacent to a disc is a positive test for assimilation. Fermentation tests are conducted by inoculating tubes of broth containing different sugars with a drop of the test yeast. Each tube also contains a small inverted glass tube (Durham tube) to detect gas production. Following incubation the presence of gas (CO_2) in the tube constitutes a positive test for fermentation. The presence of yeast sediment or a change in color of the pH indicator is not, in itself, indicative of fermentation. The biochemical pathway

Figure 19.2 Drawings of vegetative cell morphology of some representative yeasts. (*a*) *Saccharomyces cerevisiae,* round to oval cells; (*b*) *Candida* sp., elongate to oval cells with buds that elongate forming a false mycelium; (*c*) *Selenotila intestinalis,* lenticular cells; (*d*) *Trigonopsis variablis,* triangular cells.

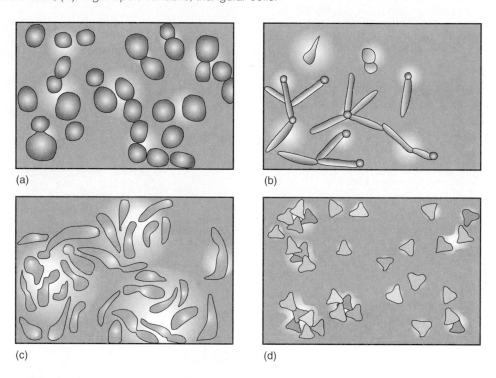

(a)

(b)

(c)

(d)

Figure 19.3 Intact asexual reproductive structure of the zygomycete *Rhizopus nigricans.* Note nonseptate coenocytic hyphae and sporangiophore.

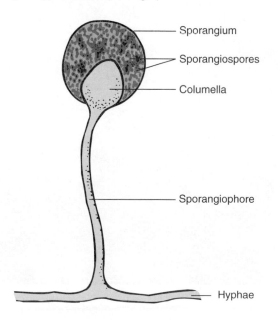

- Sporangium
- Sporangiospores
- Columella
- Sporangiophore
- Hyphae

Figure 19.4 Intact asexual reproductive structure of the ascomycete *Aspergillus niger.* Note the presence of a foot cell, a columella, a septate conidiophore, and hyphae.

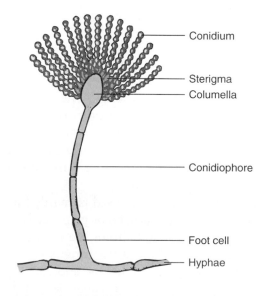

- Conidium
- Sterigma
- Columella
- Conidiophore
- Foot cell
- Hyphae

Figure 19.5 Intact asexual reproductive structure of the ascomycete *Penicillium.* Note the absence of a columella and a foot cell. Also note the symmetrical attachment of the metulae to the conidiophore—an important diagnostic feature for differentiation within the genus. Also note septate hyphae.

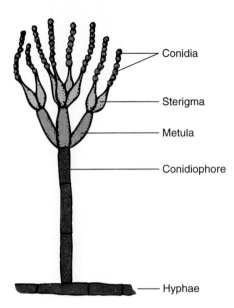

— Conidia

— Sterigma

— Metula

— Conidiophore

— Hyphae

Figure 19.6 *Schizosaccharomyces pombe,* one of two yeasts known to multiply vegetatively by fission. Courtesy of Dr. David Yarrow, The Central Bureau for Fungus Cultures, Baarn, Holland.

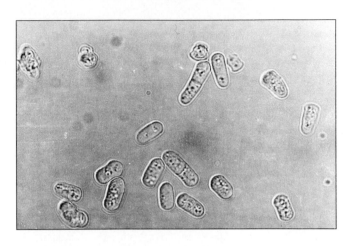

Figure 19.7 Sexual asci and ascospores of *Saccharomyces cerevisiae* showing asci with four or perhaps fewer spores. Courtesy of University of Washington.

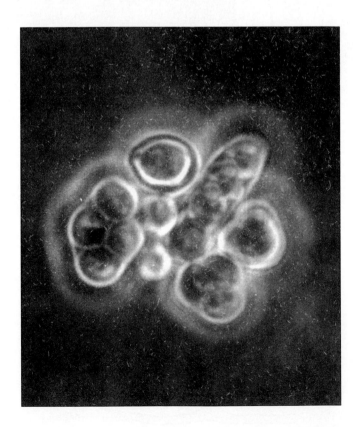

for fermentation of sugar into alcohol, CO_2, and other end products, the Embden-Meyerhof pathway **(glycolytic pathway),** is discussed in Nester, et al. p. 146.

The media used for initial isolation of many fungi is malt extract agar for the yeasts and Sabouraud's dextrose agar, with or without antibiotics to inhibit the growth of contaminant bacteria, for the filamentous fungi. The temperature of incubation depends on the organisms sought (20° to 25°C is suitable for most yeasts and filamentous fungi). One exception is *Aspergillus fumigatus* which grows well at 45°C, a temperature that inhibits growth of most other fungi.

Petri dish cultures are used primarily for study of colonial morphology (see color plates 2, 12, and 15), whereas covered slide cultures are used for detailed study of intact filamentous fungus reproductive structures (see figures 19.3, 19.4, and 19.5). They are preferred for this purpose because petri dish culture growth is often too dense to see individual, intact fruiting structures. See Appendix 7 for a method to make covered slide cultures. Also when a wet mount

Table 19.2   Some Important Pathogenic Yeasts or Yeastlike (Dimorphic) Organisms

Organism	Morphology	Ecology & Epidemiology	Diseases	Treatment
Cryptococcus neoformans	Single budding cells, encapsulated	Found in soil and pigeons' nests. No transmission between humans and animals. May be opportunistic.	Meningitis, pneumonia, skin infections, visceral organs	Amphotericin B
Candida albicans	Budding cells, pseudomycelium formation chlamydospores	Normal inhabitants of mouth, intestinal tract. Opportunistic infections.	Thrush, vaginitis, nails, eyes, lungs, systemic infections	Alkaline mouth and douche washes; parahydroxy-benzoic acid esters; amphotericin B
Blastomyces dermatitidis	37°C: single large budding cells, 20°C: mold with conidia	Disease of North America and Africa. Found occasionally in nature. No transmission between humans and animals.	Primarily lungs. Also skin and bones	High-calorie, high-vitamin diet; bed rest; aromatic diamidines; amphotericin B
Paracoccidioides braziliensis (Blastomyces braziliensis)	37°C: single and multiple budding cells 20°C: mold with white aerial mycelium	Confined to South America. Workers in close association with farming.	Chronic granulomatous infection of mucous membranes of mouth, adjacent skin, lymph nodes, viscera	Sulfonamides; amphotericin B
Histoplasma capsulatum	37°C: single small budding cells 20°C: mold with tuberculate chlamydospores	Saprophyte in soil. No transmission between humans and animals. Epidemics from silos, chicken houses, caves, etc.	Primarily lungs, may spread to reticuloendothelial system	Amphotericin B
Coccidioides immitis	37°C: thick-walled endospore filled, spherical cells 20°C: mold with arthrospores	Disease primarily of arid regions, such as San Joaquin valley. Dust-borne disease. No transmission between humans and animals.	Primarily lungs may disseminate particularly in African-Americans and is highly fatal	Bed rest; amphotericin B; surgery for lung lesions

is prepared from a petri dish culture, the intact fruiting structures are usually broken apart, leaving only individual parts of the fruiting structure. Wet mounts prepared from petri dish cultures are useful in making detailed microscopic observations of individual parts of the filamentous fungus, e.g., the sporangium, conidium, etc. (see figures 19.3, 19.4, and 19.5) where these structures are labeled.

Many of the medically important fungi are found in the classes Ascomycetes and Deuteromycetes (see table 19.1). Most of their infections are opportunist limited to cutaneous or subcutaneous tissues. Such infections can sometimes become progressive leading to systemic involvement with the possibility of death. According to Al-Doory, the use of new medical technologies such as prolonged or extensive use of antibiotics, anticancer agents, and immunosuppressive drugs in organ transplants is expected to continue, thus increasing the ever present risk of opportunist fungal infections. If so, there will be an expanded need for trained mycologists and clinical mycology laboratories. Clinically there are three types of such **mycoses:**

1. **Dermatomycoses** are superficial **keratinized** infections of the skin, hair, and nails caused by a group of filamentous fungi commonly called dermatophytes. They rarely invade

Figure 19.8 Chlamydospores and smaller blastospores, (attached to pseudohyphae) of *Candida albicans* grown on cornmeal agar. Preparation stained with methylene blue.
Courtesy of the Upjohn Co.

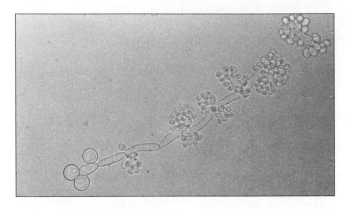

Figure 19.9 Germ tubes formed by *Candida albicans* grown on egg albumin. Phase-contrast magnification.
Courtesy of the Upjohn Co.

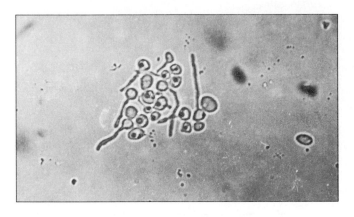

subcutaneous tissues. They show rudimentary morphology, appearing only as mycelial growth on skin and nails, or as fragments of **mycelium** and **arthrospores** arranged inside and outside of hair. In all instances, they form circular lesions described as **ringworm** (see color plate 10). However, in culture they form filamentous colonies and asexual reproductive spores.

2. **Subcutaneous mycoses** are caused by either filamentous or by **dimorphic** yeastlike fungi (see color plates 11 through 14). They also remain fixed at the site of infection.

3. **Systemic yeast** and **yeastlike infections** originate in the lungs, and can spread to other parts of the body. Examples include **histoplasmosis**, which is caused by the dimorphic yeastlike fungus *Histoplasma capsulatum*, and **coccidioidomycosis** (San Joaquin fever) caused by another yeastlike fungus, *Coccidioides immitis* (see color plates 15 through 17) and exercises 30 and 31 where *C. immitis* is diagnosed in humans using two widely used serological tests: the ELISA test and the Ouchterlony test.

Definitions

Arthrospores.   Thick-walled asexual spores formed by breaking apart of septate hyphae.

Ascospore.   Sexual spore characteristic of the fungus class Ascomycetes.

Ascus.   Saclike structure containing ascospores.

Assimilation.   Ability, in the presence of oxygen, to utilize carbohydrates for growth.

Blastospore.   Asexual spore formed by budding from a cell or from hyphae.

Budding.   An asexual process of reproduction in which a daughter cell (bud) evolves from either a larger cell (mother cell) or from a hyphae.

Chlamydospore.   A resistant hyphal cell with a thick wall; it eventually separates from the hyphae and functions as a spore.

Coccidioidomycosis.   An upper respiratory tract infection caused by the dimorphic yeastlike organism, *Coccidioides immitis*.

Coenocytic.   A multinuclear mass of protoplasm resulting from repeated nuclear division unaccompanied by cell division.

Columella.   A swelling of the sporangiophore at the base of the sporangium, which acts as a support structure for the sporangium and its contents.

Conidia.   Asexual spores produced from either the tip or side of the conidiophore, or hypha.

Culture spherule. A thick-walled sphere-shaped cell containing many small endospores, characteristic of the tissue phase of *Coccidioides immitis*.

Cyst. A resting (dormant) spore.

Daughter cell. A new yeast cell. Also see budding.

Dermatomycosis. A disease of the skin caused by infection with a fungus.

Dimorphic. Ability to exist in two forms, e.g., in the fungi either a mycelial or yeastlike form.

Fermentation. Growth in the absence of oxygen in which the final electron acceptor is an organic compound.

Fission. An asexual process in which one cell splits into two or more daughter cells.

Foot cell. A cell located at the base of the conidiophore in the genus *Aspergillus*.

Germ tube. A tubelike outgrowth from an asexual yeast cell that develops into a hypha.

Glycolytic pathway. An initial series of fermentation steps in which carbohydrates are degraded. Often called the Embden-Meyerhof pathway.

Histoplasmosis. A pulmonary infection caused by *Histoplasma capsulatum*, a dimorphic yeast.

Hypha (pl., hyphae). Threadlike fungal filament(s) that form a mycelium.

Keratinophilic. The ability of certain dermatophytic fungi to utilize a highly insoluble body protein called keratin. Examples include skin, hair, and nails.

Lichen. A symbiotic relationship between a fungus and an alga. See Nester et al. for additional information.

Metula. A branch(s) at the tip of the conidiophore which supports sterigmata.

Mold. A filamentous fungus often appearing as woolly growth on decaying materials.

Mycelium. A fungal mat made of tangled hyphae.

Mycoses. Diseases caused by fungi.

Nonfilamentous fungi. Fungi devoid of hyphae, e.g., yeast.

Opportunist. An organism capable of causing disease only when host defense mechanisms are impaired.

Rhizoids. Rootlike structures made of fungus hyphae that are able to penetrate various substrates in order to anchor the fungus so that it can obtain nutrients.

Ringworm. Contagious fungal diseases of the hair, skin, or nails. See color plate 10.

Saprophyte. An organism that obtains nourishment from decayed organic matter.

Spherule. A large thick-walled structure filled with fungal endospores. See color plate 17.

Sporangiospore. A sexual reproductive spore found in the Zygomycetes.

Sterigma (pl, sterigmata). A specialized hypha that supports either a conidiospore(s) or a basidiospore(s).

Stolon. A runner, such as found in strawberry plants, made of horizontal hyphae from which sporangiospores and rhizoids originate. Stolons are characteristic of the class Zygomycetes.

Systemic yeast. Yeast found in various parts of the body.

Toadstool. A large filamentous fleshy fungus with an umbrella-shaped cap. See color plate 8.

Yeast. A nonfilamentous fungus often found in nature on fermenting fruits and grains.

Yeast dimorphism. Existing in two growth forms, such as the mold phase (hyphal filaments) and yeast phase (single cells) of pathogenic fungi.

Objectives

1. To introduce information in the Getting Started section about what fungi are, and how to distinguish them from one another, as well as how to identify members of the two major groups of fungi, the nonfilamentous and the filamentous fungi.

2. Included for identification studies of the nonfilamentous fungi (yeasts) are two members of the class Ascomycetes: *Saccharomyces cerevisiae* and a dimorphic

yeast, *Candida albicans*. Morphological studies to be used for their identification include colonial and vegetative cell morphology, sexual ascus and ascospore morphology, and chlamydospore, germ tube, and pseudohyphae formation. The latter three studies represent examples of yeast dimorphism for the genus *Candida albicans*.

3. Included for studies of the filamentous fungi is a member of the class Zygomycetes, *Rhizopus nigricans* and two members of the class Deuteromycetes, *Aspergillus niger* and a *Penicillium* species. Morphological studies to be used for their identification are macroscopic and microscopic studies of colonial and vegetative cell morphology when cultured on agar plates and perhaps when cultured using a covered slide culture.

References

Al-Doory, Y. *Laboratory medical mycology.* Philadelphia: Lea and Febiger, 1980.

Barnett, H.L. *Illustrated genera of imperfect fungi,* 2nd edition, fifth printing. Burgess Publishing Company, 1969. Collection of drawings describing 302 genera of Fungi Imperfecti.

Barnett, J.A.; Payne, R.W.; and Yarrow, D. *Yeasts: Characteristics and identification,* 2nd ed. New York: Cambridge University Press, 1991.

Larone, D. *Medically important fungi, a guide to identification,* 2nd ed. Washington, D.C.: American Soc. for Microbiology.

Moore, D. *Fungal morphogenesis.* Cambridge University Press, 1998.

Nester et al. *Microbiology: A human perspective,* 4th ed., 2004. Chapter 12, Section 12.3.

Phaff, H.H.; Miller, M.W.; and Mrak, E.M. *The life of yeasts,* 2nd ed. Cambridge, Mass.: Harvard University Press, 1978. This book provides an excellent introduction to yeast morphology, propagation, their cytology, ecology (where and how they propagate in nature), and their role as food spoilage organisms, as well as their use in various industries.

Rippon, J.W. *Medical mycology,* 3rd ed. Philadelphia: W.B. Saunders Co., 1988.

Materials

Per team of three students

Cultures

Sabouraud's dextrose broth cultures (48 hr, 25°C) of *Saccharomyces cerevisiae* and *Candida albicans*. Note: *Candida albicans* is a potential pathogen, especially with some immunocompromised individuals. Such individuals may wish to consult beforehand with their physician.

Sabouraud's dextrose agar slant culture (48 hr, 25°C) of *Candida albicans*.

Sabouraud's dextrose agar petri dish cultures (3–5 days, 25°C) of *Rhizopus nigricans*, *Aspergillus niger*, and *Penicillium notatum*.

Perhaps covered slide cultures of the three above filamentous fungi on Sabouraud's dextrose agar (3–5 days, 25°C). See Appendix 7 for instructions to prepare such a culture. You may wish to consider it as a special project exercise.

Tubes of glucose, maltose, and lactose broth containing Durham tubes, 2 tubes of each

Glucose-acetate yeast sporulation agar, 1 plate

Cornmeal agar, 1 plate

Test tube (12 by 75 mm) containing either 0.5 ml of serum or raw, nonsterile egg white, 1

Sterile droppers, 4

Tweezers

Dissecting microscope(s)

Ruler divided in mm

Dropping bottle containing methylene blue

Procedure

First Session

Suspend the broth cultures of *Saccharomyces cerevisiae* (S.c.) and *Candida albicans* (C.a.).

1. Yeast fermentation study. Inoculate each of the carbohydrate fermentation tubes (glucose, lactose, and maltose) with a loopful of *S.c.* Repeat using fresh tubes with *C.a.* Place the properly labeled tubes in a container and incubate at 25° to 30°C for 48 hours.

2. Yeast colonial and vegetative cell morphology study. Divide the bottom surface of the Sabouraud's dextrose agar plate in half with a marking pencil. Label one half *S.c.* for *Saccharomyces cerevisiae* and the other half *C.a.* for *Candida albicans*. With a sterile dropper, inoculate the agar surface of each sector with a <u>small</u> drop of the respective test yeast. Allow the inoculum to soak into the agar before incubating right side up in the 25° to 30°C incubator for 48 hours.

3. Yeast sexual sporulation study. With a sterile dropper inoculate the center of the sporulation agar plate with a small drop of the *S.c.* broth culture. Allow the inoculum to soak into the agar before incubating right side up in the 25° to 30°C incubator for 48 or more hours. Cultures freshly isolated from nature generally sporulate much faster than laboratory held cultures.

4. Yeast chlamydospore formation study. Inoculate the center of a cornmeal agar plate with a small drop of the *C.a.* broth culture. With a sterile loop, streak the drop across the length of the plate with just enough pressure to mark but not cut the agar. Next, streak back and forth across the marked area. Grasp a coverslip with a pair of sterile tweezers (sterilize by dipping in alcohol and passing through the Bunsen burner flame) and then place the coverslip over a portion of the streaks. Incubate the plate right side up in the 25° to 30°C incubator for two to four days.

5. Yeast germ tube formation study.
   a. Preparation. Remove a loopful of *C.a.* from the surface of the Sabouraud's dextrose agar slant culture. Emulsify the inoculum in the tube of serum or raw egg white. Incubate the tube for 2½ to 3 hours in the 37°C incubator.
   b. Observation. Mix the serum or raw egg white with a dropper, and prepare a wet mount using a single drop of the mixed suspension.

Examine first with the low power objective and next with the high power objective. Look for short germ tubes that give the cells a drumstick appearance (see figure 19.9).

Prepare drawings of your findings in the part 5 Results section of the Laboratory Report.

Note: If insufficient time remains for observation, the tubes can be held until the Second Session for observation by storing them in a covered container in the refrigerator.

6. Colonial characteristics of petri dish cultures of *Rhizopus nigricans* (*R.n.*), *Aspergillus niger* (*A.n.*), and *Penicillium notatum* or other species of *Penicillium* (*P.sp.*).

   Visually examine each petri dish culture noting the following:
   a. Colony size. With a ruler measure the diameter in mm.
   b. Colony color. Examine both the upper and lower surfaces.
   c. Presence of soluble pigments in the agar medium.
   d. Colony texture (such as cottony, powdery, or woolly).
   e. Colony edge (margin). Is it regular or irregular?
   f. Colony convolutions (ridges). Are they present?

   Enter your findings in table 19.5 of the Laboratory Report.

7. Morphological study of asexual fruiting structures found in *Rhizopus*, *Aspergillus*, and *Penicillium* species.

   The best way to make studies of this type is with a covered slide culture. In the event none is available, you can attempt to do so with your petri dish culture. A major problem is the density of growth in the petri dish culture which makes it difficult to find intact asexual reproductive structures. However they can often be found for *Rhizopus nigricans*, in that, like strawberries, it has **stolons** which enable it to spread and attach to the underside of the petri dish lid. A dissecting microscope is preferable for use in making your initial observations. If not available, the light microscope, with the low power objective, can also be used. The low power objective more than doubles the magnification obtained with the dissecting microscope. Procedurally do as follows:

Place the covered petri dish culture on the stage of either the dissecting microscope or the light microscope, and examine a sparsely populated area of the colony for the presence of asexual reproductive structures (see figure relating to the fungus culture being examined, either 19.3, 19.4, or 19.5).

Note: Never smell fungus cultures—spore inhalation may cause infection. When you first observe fruiting bodies, stop moving the petri dish, and keeping the air currents to a minimum to avoid spore dispersal, carefully remove the petri dish cover, and re-examine to determine whether you can in fact see the various parts of the fruiting body as described in the figure for that fungus. In the case of R.n. you should be able to see fruiting bodies by examining the underside of the petri dish lid. It may take as long as 4 to 5 days incubation before finding *Rhizopus* fruiting bodies with stolons. If intact fruiting bodies are found for any of the three filamentous fungi, make drawings of their asexual reproductive structures in part 7 of the Laboratory Report. Label the parts in a manner similar to that used in figures 19.3, 19.4, and 19.5. Covered slide cultures are the answer if fruiting bodies cannot be found for the A.n. and P.sp.

8. Detailed examination of **sporangiospores, conidia,** and if present, **chlamydospores:**

For observing sporangiospores (R.n.) and conidia (A.n. and P.sp.) remove some aerial growth with a loop. Prepare a wet mount and observe it with the low and high dry objectives of the microscope.

Chlamydospores can be found in both surface and submerged R.n. mycelium. They are elongated, brown in color and have thick walls. Prepare a wet mount and observe with both the low and high dry power objectives.

Note: You may wish to first search for spores in both the inner and outer fringes of the colony using the low power objective. To better search the inner colony surface, make the area less dense by first removing some of the aerial growth with a loop. Flame the loop to destroy the spores. Prepare and label drawings of the various asexual spore types found, in part 8 of the Laboratory Report. Some morphological characteristics of value for identification are:

1. *Rhizopus nigricans*
   a. Has nonseptate coenocytic hyphae
   b. Contains **rhizoids.** See underside of the petri dish of an older culture.
   c. Details of fruiting body (see figure 19.3). Note the nonseptate stem (sporangiophore), the swelling at the tip of the sporangiophore (**columella),** and the sac that encloses the columella (**sporangium),** which contains the asexual reproductive spores (sporangiospores).

2. and 3. *Aspergillus niger* and *Penicillium notatum*
   These molds have fruiting bodies somewhat similar in appearance. Both have brushlike structures comprised of bottle-shaped cells (**sterigmata)** to which are attached long chains of asexual reproductive spores (**conidiospores).** They differ in that the genus *Aspergillus* has a swollen cell at the base of the stem (**conidiophore)** known as the **foot cell,** as well as a columella to which the sterigmata are attached (see figure 19.4). The sterigmata may occur in one or two series depending upon the species involved. Finally in the genus *Penicillium* the branching at the tip of the conidiophore can be either symmetrical (see figure 19.5) or asymmetrical, depending on the point of attachment of the **metulae** to the conidiophore. When all of the metulae are attached at the tip of the conidiophore, branching of the sterigmata and attached conidia will appear symmetrical (see figure 19.5). If one of the metulae in figure 19.5 is attached below the tip of the conidiophore, an asymmetrical branching occurs. This is an important diagnostic feature for differentiation within the genus *Penicillium*.

Second Session

1. Yeast fermentation study. Examine the fermentation tubes for the following and record your results in table 19.3 of the Laboratory Report:
   a. Presence or absence (+ or −) of cloudy broth (growth).
   b. Presence or absence (+ or −) of gas in the inverted Durham tube.
   c. Change in color of the pH indicator dye. A change to a yellow color is indicative of acid production.

   Note: Gas production is indicative of fermentation (glycolysis). To detect false negative results caused by super saturation of the broth, all tubes giving an acid reaction should be shaken lightly and the cap vented. This operation is frequently followed by a rapid release of gas. All positive fermentation reactions with a carbohydrate sugar are accompanied by positive assimilation of that carbohydrate as evidenced by increased clouding of the broth; however, sugars may be assimilated without being fermented.

2. Yeast colonial and vegetative cell morphology study.
   a. Colony characteristics. If possible observe the Sabouraud's agar plate over a 5–7 day incubation period. Make note of the following in table 19.4 of the Laboratory Report: colony color; consistency (soft, firm), probe the colony with a sterile needle for this determination; colony diameter (mm); colony surface (rough or smooth, flat or raised); and appearance of the colony edge (circular or indented).
   b. Vegetative cell morphology. Remove a loopful of surface growth from each colony and prepare wet mounts. Observe with the high dry objective, noting the shape and size of the cells and the presence or absence of pseudohyphae (see figure 19.2b). Prepare and label drawings of the two yeasts in part 2b of the Laboratory Report.

3. Sexual sporulation study. With a sterile loop, touch the *S. cerevisiae* colony on the glucose-acetate agar plate and prepare a wet mount. Observe with the high dry objective and look for the presence of asci containing 1 to 4 or perhaps more ascospores (see figure 19.7). Prepare and label drawings of your findings in part 3 of the Laboratory Report.

   Note: In the event you do not find asci, reincubate the plate up to one week, perhaps even longer, and reexamine periodically. Some yeast strains take longer than others to produce sexual spores.

4. Chlamydospore formation study. Remove the cover of the cornmeal agar plate and place the plate on the microscope stage; with the 10× objective focus on the edge of the coverslip and search for chlamydospores that are quite large (7 to 17 um). When present, they will usually be found underneath the coverslip near the edges. In addition you should find thin pseudohyphae, and many very small round blastospores. Prepare and label drawings of your findings in part 4 of the Laboratory Report (see figure 19.8). You can also remove some of the agar with a loop from the marked area, place it on a slide, and prepare a wet mount with a drop of methylene blue, a dye capable of staining the chlamydospores. The slide can be observed with both the low and high dry objectives.

5. Germ tube formation study. Previously discussed in the First Session.

6., 7., and 8. Filamentous fungi. Complete any remaining morphological studies.

   Note: If an ocular micrometer is available and time permits you may wish to make measurements of some of the various morphological structures, e.g., a comparison of asexual spore sizes of different filamentous fungi. Appendix 5 contains information on use, calibration, and care of the ocular micrometer.

NOTES:

EXERCISE

Laboratory Report: Microscopic Identification of Fungi—
Nonfilamentous and Filamentous Fungi

Results (Nonfilamentous Fungi)

1. Fermentation Study. Examine tubes and record results (+ or −) in table 19.3. For details see Procedure, Second Session step 1, p. 161.

Table 19.3   *C. albicans* and *S. cerevisiae* Fermentation Activity in Tubes of Broth Containing Different Carbohydrate Sugars

Yeast Strain	Glucose			Maltose			Lactose		
	Cloudy	Gas	Acid	Cloudy	Gas	Acid	Cloudy	Gas	Acid
C. albicans									
S. cerevisiae									

2. Yeast Colonial and Vegetative Cell Morphology Study.
   a. Colony characteristics (see Procedure, Second Session step 2a, p. 161) and enter results in table 19.4.

Table 19.4   *C. albicans* and *S. cerevisiae* Colonial Characteristics on Sabouraud's Dextrose Agar Plates

Yeast Strain	COLONIAL MORPHOLOGY				
	Colony Color	Consistency	Diameter (mm)	Surface Appearance	Edge Appearance
Candida albicans					
Saccharomyces cerevisiae					

b. Vegetative cell morphology (see Procedure, Second Session step 2b, p. 161) and enter results below:
    Candida albicans

    Saccharomyces cerevisiae

3. Sexual Sporulation Study (*S. cerevisiae*). Drawings of asci and ascospores (see Procedure, Second Session step 3, p. 161).

4. Chlamydospore Formation Study (*C. albicans*). Drawings of chlamydospores (see Procedure, Second Session step 4, p. 161).

5. Germ Tube Formation Study (*C. albicans*). Drawings of germ tubes (see Procedure, First Session step 5, p. 159).

6. Colonial characteristics of *Rhizopus*, *Aspergillus*, and *Penicillium* when grown on Sabouraud's dextrose agar. Describe in table 19.5.

Table 19.5   Colonial Characteristics of Three Filamentous Fungi Cultured for _____ Days on Sabouraud's Dextrose Agar

	Rhizopus	*Aspergillus*	*Penicillium*
Colony color			
Colony diameter (mm)			
Colony texture			
Colony convolutions			
Colony margin			
Soluble pigments in agar			

7. Drawings of their asexual reproductive structures (please label parts).

Rhizopus	*Aspergillus*	*Penicillium*

8. Drawings of their asexual spores (please label parts).

     Rhizopus   **Aspergillus**   **Penicillium**

Questions

1. List four ways of differentiating *Candida albicans* from *Saccharomyces cerevisiae*.

2. What are two ways in which you were able to differentiate pathogenic from nonpathogenic *Candida* species?

3. Explain the physiological differences between yeast fermentation and yeast assimilation of glucose.

4. Why is a loop rather than a pipet used to inoculate the sugar fermentation tubes?

5. Why would the growth of a pellicle or film on the surface of a broth growth medium be advantageous to the physiology and viability of that yeast?

6. What are some ways in which you might be able to differentiate *Rhizopus nigricans* from *Aspergillus niger* simply by visually observing a petri dish culture?

7. How can you determine whether or not a green, woolly looking colony is an *Aspergillus* or *Penicillium*?

8. What problems might you have in identifying a pathogenic fungus observed in a blood specimen? What might you do to correct such problems?

9. In what ways can we readily distinguish:
   a. fungi from algae?

   b. fungi from bacteria?

   c. fungi from actinomycetes?

10. Define an opportunistic fungus. Provide some examples. Are all medically important fungi opportunistic? Feel free to discuss your answers.

11. Name three pathogenic fungi that exhibit dimorphism. Describe the type of dimorphism each exhibits and the laboratory conditions necessary to elicit it.

EXERCISE

Parasitology: Protozoa and Helminths

Getting Started

Since the natural histories of parasitic diseases differ in some important respects from those of bacterial diseases, they merit a separate laboratory session to give you introductory laboratory experience with parasites, the diseases they cause, and techniques used to diagnose them.

The distinguishing feature of parasitic life is the close contact of the parasite with the host in or on which it lives, and its dependency on the host for life itself. This special association has led to the evolution of three types of adaptations not found in the free-living relatives of the parasites: loss of competency, special structures, and ecological ingenuity.

Parasites have become so dependent on their hosts for food and habitat that they now experience a *loss of competency* to live independently. They usually require a specific host and many have lost their sensory and digestive functions; these are no longer important for their survival.

On the other hand, they have developed *special structures* and functions not possessed by their free-living relatives, which promote survival within the host. One example is special organs of attachment—hooklets and suckers. Parasites also have a tremendously increased reproductive capacity, which compensates for the uncertainty in finding a new host. Tapeworms, for example, have fantastically high rates of egg production, reaching up to 100,000 per day.

Ecological ingenuity is demonstrated in the fascinating variety of infecting and transmitting mechanisms. This has led to very complex life cycles, which contrast markedly with the relatively simple lifestyles of their free-living counterparts. Parasites show quite a range in the types of life cycles they possess, from species that pass part of each generation in the free-living state to others that require at least three different hosts to complete the life cycle. Some are simply transmitted by insects from

one human host to the new host, or the insect may act as a host as well. Many protozoa develop resistant cysts that enable them to survive in unfavorable environments until they find a new host. The eggs of flatworms and roundworms also have a protective coat.

These three strategies promote survival and expansion of the species by providing greater opportunities for finding and infecting new hosts, which is a continual problem for parasites. Successful interruption of these cycles to prevent their completion is an important feature of public health measures used to control diseases caused by parasites.

This exercise is designed to give you some practical experience with representative protozoan and helminthic parasites, and with clinical methods used in their diagnosis and control. Your study will consist of these procedures:

1. As an introduction, you will have an opportunity to observe the movements and structure of some living nonparasitic protozoans and worms often found in pond water.
2. Examination of commercially prepared stained blood and fecal slides that contain human protozoan parasites.
3. Microscopic comparison of the structure of parasitic worms with that of their free-living relatives to observe some special adaptations to the parasitic way of life.
4. Study of the natural history and life cycle of the human parasitic disease schistosomiasis. This will enable you to see the interaction between stages of the life cycle, environmental surroundings, and social conditions of their human hosts as factors in the epidemiology and control of the disease.

The following classification of parasites will serve as a guide to the examples you will be studying in this exercise. It is not a complete listing.

Protozoa

Protozoa, a subkingdom of the kingdom Protista, are unicellular eukaryotic organisms. They usually reproduce by cell division and are classified into five phyla mainly according to their means of locomotion. Only one of these phyla, the Phylum Suctoria, which is closely related to the Phylum Ciliata, does not contain animal pathogens. The remaining phyla are classified as follows:

Phylum Sarcodina

Members of this phylum move and feed slowly by forming cytoplasmic projections known as **pseudopodia** (false feet). They also form both **trophozoites** (vegetative form) and **cysts** (resistant, resting cells). Parasitic members include the **amoeba** *Entamoeba histolytica*, which causes amoebic dysentery. It ingests red blood cells and forms a four-nucleate cyst. It is also found in animals. Other amoeba species found in humans, such as *Entamoeba gingivalis*, are relatively harmless **commensals.**

Phylum Ciliata

Members of this phylum have many short, hairlike cilia on their body surfaces that beat rhythmically by bending to one side. They contain two nuclei: a macronucleus and a micronucleus. This phylum is typified by the genus *Paramecium* (see color plate 18), a nonpathogenic form readily found in pond water. Another member, *Balantidium coli*, is a common parasite in swine, and can infect humans, causing serious results.

Phylum Mastigophora

These protozoans propel themselves with one or more long, whiplike flagella. Some have more than one nucleus and usually produce cysts. Different species cause infections in the intestines, vagina, blood, and tissues. *Giardia lamblia* (see figure 33.3) causes a mild to severe diarrheal infection. *Trichomonas vaginalis* (figure 20.1) is found in the urogenital region, where it causes a mild vaginitis in women. *Trypanosoma gambiense* (figure 20.2) infects the blood via tsetse fly bites, where it causes trypanosomiasis, or African sleeping sickness, in cattle and humans. Cattle and other ungulates serve as a reservoir for this organism.

Figure 20.1 *Trichomonas vaginalis.* Illustration of a typical mastigophoran protozoan. From Eugene W. Nester et al. *Microbiology: A Human Perspective.* WCB McGraw-Hill. 2001. All Rights Reserved. Reprinted by permission.

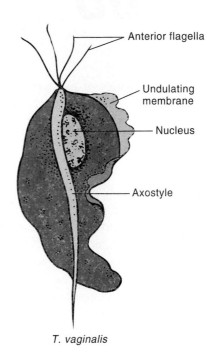

T. vaginalis

Figure 20.2 A prepared blood smear of trypanosome parasites in human blood. The slender flagellates lie between the red blood cells. © Cleveland P. Hickman, Jr./Visuals Unlimited

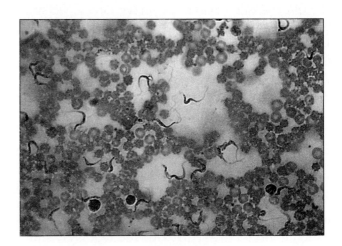

Phylum Sporozoa

These are obligate, nonmotile parasites with alternating stages: the sexual reproductive stage is passed in the definitive insect host and the asexual phase in the intermediate human or animal host. The genus *Plasmodium* includes the malarial species, in which the **definitive host** is the female *Anopheles* mosquito, and the **intermediate host** humans. The genus *Coccidia* includes important intestinal parasites of fowl, cats, dogs, swine, sheep, and cattle. *Toxoplasma* species is a cat parasite that can harm the human fetus in an infected pregnant woman.

Helminths (Worms)

These are multicellular eukaryotic organisms. Two of the phyla, Platyhelminthes (flatworms) and Nemathelminthes (roundworms) contain pathogenic worms.

Phylum Platyhelminthes

Members of the Platyhelminthes are flat, elongated, legless worms that are **acoelomate** and exhibit bilateral symmetry. This phylum contains three classes:

Class Turbellaria

These are free-living **planarians** (flatworms), such as are found in the genus *Dugesia* (figure 20.3).

Class Trematoda (Flukes)

They have an unsegmented body and many have suckers to hold them onto the host's intestinal wall.

Figure 20.3 The genus *Dugesia,* a free-living planarian in the class Turbellaria. (*a*) A living specimen (× 0.5) is shown, and (*b*) shows a labeled line drawing. (*a*) © John D. Cunningham/Visuals Unlimited

(a)

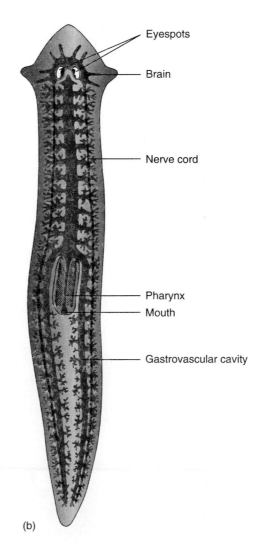

- Eyespots
- Brain
- Nerve cord
- Pharynx
- Mouth
- Gastrovascular cavity

(b)

Many flukes have complex life cycles that require aquatic animal hosts. The *Schistosoma* species are bisexual trematodes that cause serious human disease. They require polluted water, snails, and contact with human skin for completion of their life cycles (see color plate 19). *Clonorchis sinensis* and *Fasciola* species are liver flukes acquired by eating infected raw fish and contaminated vegetables, respectively.

Class Cestoda (Tapeworms)

These are long, segmented worms with a small head (**scolex**) equipped with suckers and often hooklets (figure 20.4) for attachment to the host's intestinal wall. The series of segments, or **proglottids,** contain the reproductive organs and thousands of eggs. These segments break off and are eliminated in the feces, leaving the attached scolex to produce more proglottids with more eggs. Figure 20.5 illustrates the life cycle of the tapeworm in a human. The symptoms of *Taenia* tapeworm are usually not serious, causing only mild intestinal symptoms and loss of nutrition. Not so for the *Echinococcus* tapeworm, which causes a serious disease. All tapeworm diseases are transmitted by animals.

Figure 20.4 Illustration of a tapeworm scolex showing both hooklets and suckers for attachment to the intestine. *T. saginata* (beef tapeworm) is essentially without hooklets, whereas *T. solium* (pork tapeworm) has both. © Stanley Flegler/Visuals Unlimited

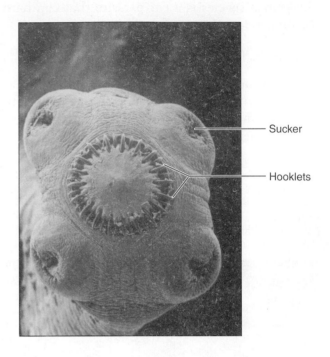

Sucker

Hooklets

Figure 20.5 Life cycle of *Taenia saginata*. The adult tapeworm with scolex and proglottids is conceived from larvae in the human intestine.

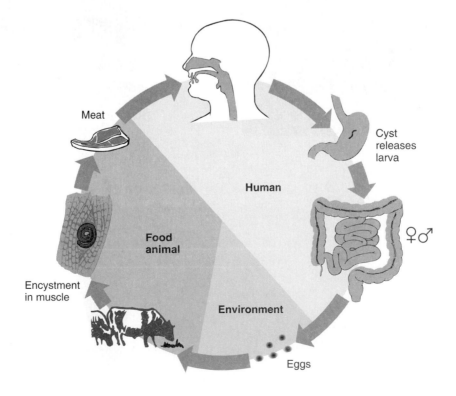

Meat

Cyst releases larva

Human

Food animal

♀♂

Encystment in muscle

Environment

Eggs

Phylum Nemathelminthes

Members of the Phylum Nemathelminthes (roundworms) occupy an important ecological niche since they are present in large numbers in very diverse environments, including soil, fresh water, and seawater. In contrast to the Platyhelminthes, these round, unsegmented worms are **coelomate** (have a body cavity), and have a complete digestive tract and separate sexes. This phylum contains many agents of animal, plant, and human parasitic diseases. Most require only one host, and can pass part of their life cycle as free-living larvae in the soil. *Trichinella spiralis* requires alternate vertebrate hosts. Humans become infected when they ingest inadequately cooked meat, such as pork or wild game, containing the larval forms in the muscles. *Ascaris lumbricoides* (figure 20.6) is probably the most common worldwide of all the human helminths. *Enterobius vermicularis* causes pinworm, a very common condition in children in the United States. Efforts to eradicate it have not been very successful since pinworm causes little discomfort. *Oxyuris* causes a similar condition in animals.

Definitions

Acoelomate. Without a true body cavity. Typical of members of the Phylum Platyhelminthes (flatworms).

Amoeba. Unicellular organisms with an indefinite changeable form.

Cercaria. The last miracidium stage in which the larvae possess a tail.

Coelomate. With a true body cavity. Typical of members of the Phylum Nemathelminthes (roundworms).

Commensal. A relationship between two organisms in which one partner benefits from the association and the other is unaffected.

Cysts. Dormant, thick-walled vegetative cells.

Definitive host. The host in which the sexual reproduction of a parasite takes place.

Intermediate host. The host that is normally used by a parasite in the course of its life cycle, and in which it multiplies asexually but not sexually.

Figure 20.6 *Ascaris lumbricoides,* an intestinal roundworm up to 12 inches long. A mass of worms recovered from the ileum of a malnourished child. From Rubin and Farber, *Pathology.* Reprinted by permission of J. B. Lippincott Company.

Merozoites. Schizont nuclei that become surrounded by cytoplasm and bud off as daughter cells or merozoites.

Miracidium. A free-swimming ciliate larva that seeks out and penetrates a suitable intermediate snail host in which it develops into a sporocyst.

Planarian. Any flatworm of the genus *Planaria*.

Proglottids. Any of the segments of a tapeworm formed in the neck region by a process of strobilation (transverse fission).

Pseudopodia. Extensions of cytoplasm that aid in engulfing particles and functioning in motility of amoeboid cells.

Schizont. A stage in the life cycle of Sporozoa in which the nucleus of the parent cell, the schizont, undergoes repeated nuclear division without corresponding cell divisions.

Scolex. The head of a tapeworm, which is used for attaching to the host's intestinal wall.

Sporocyst. A stage in the life cycle of certain protozoa in which two or more of the parasites are enclosed within a common wall.

Trophozoites. Vegetative forms of some protozoans.

Objectives

1. To introduce you to the study of parasitology by letting you examine some examples of nonparasitic protozoans (*Amoeba proteus* and a *Paramecium* species) and a free-living planarian (flatworm) of the genus *Dugesia*.

2. To study the morphology of some free-living, trophozoite, and cystic forms of intestinal parasites using prepared slides, and malarial and trypanosome parasites using stained blood smears.

3. To study the natural history and life cycle of an important human parasitic disease, schistosomiasis, using stained slides and a life cycle diagram (figure 20.7).

4. To demonstrate special adaptations of parasitic worms through a study of stained slides in which they are compared to similar free-living relatives.

References

Leventhal, R., and Cheadle, R. *Medical parasitology*, 4th ed. F. A. Davis, Philadelphia, 1994.

Markell, E.; Voge, M.; and John, D. *Medical parasitology*, 8th ed., Saunders, Philadelphia, 1999.

Nester et al. *Microbiology: A human perspective*, 4th ed., 2004. Chapter 12, Section 12.2 and Section 12.5.

Neva, F. A., and Brown, H. W. *Basic clinical parasitology*, 6th ed., Appleton Lange, Norwalk, 1994.

Figure 20.7 *P. falciparum* infection showing the ring trophozoite stage in a blood smear. Courtesy of the Centers for Disease Control

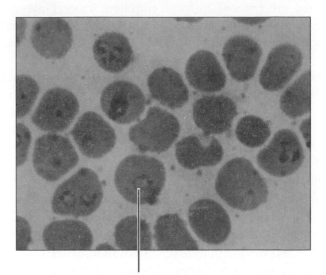

Ring trophozoite

Materials

Cultures

Living cultures of a *Paramecium* species, *Amoeba proteus*, and a *Dugesia* species

If available, a fresh sample of quiescent, stagnant pond water, which often contains members of the above genera. Students may wish to bring their own pond water.

Planaslo solution, 1 or more dropping bottles

Depression slides (hanging drop slides)

The following commercially prepared slides:

Subkingdom Protozoa

Phylum Sarcodina (pseudopodia)

Entamoeba histolytica trophozoite and cyst stages

Phylum Ciliata (cilia)

Paramecium trophozoite

Phylum Mastigophora (flagella)

Giardia lamblia trophozoite and cyst stages

Trypanosoma gambiense

Phylum Sporozoa (nonmotile)

Plasmodium vivax ring, amoeboid
schizont stages
Subkingdom Helminths (worms)
Phylum Platyhelminthes (flatworms)
Class Turbellaria (free living)
Dugesia species
Class Trematoda (flukes)
Schistosoma mansoni
adult male
adult female
ovum (egg)
ciliated miracidium
infective ciliate cercaria
sporocyst stage in snail liver tissue
Clonorchis sinensis
Class Cestoda (tapeworms)
Taenia solium trophozoite

Procedure

Note: If the number of prepared slides is limited, these procedures may be performed in a different order to facilitate sharing of slides.

1. Examination of free-living cultures:
   a. Pond water examination. Prepare a hanging drop slide (see exercise 3). Examine initially with the low power objective and later with the high and oil immersion objectives. Observe the mode of locomotion of any amoeboid or paramecium-like protozoans found. If their movements are too rapid, add a drop of Planaslo to slow them down. Describe their movements and prepare drawings in part 1a of the Laboratory Report.
   b. Examination of fresh samples of an amoeba (such as *Amoeba proteus*), paramecium (for example, *Paramecium caudatum*), and a free-living flatworm (such as *Dugesia* species). Use hanging drop slide preparations and examine as described in 1a for pond water.

Record your observations in part 1b of the Laboratory Report.

Note: Although less desirable for observing motion, wet mounts may be substituted if necessary for hanging drop slides.

2. Examination of stained slides for trophozoites and cysts:
   a. Using the oil immersion objective, examine prepared slides of a protozoan, either the amoeba *Entamoeba histolytica* or the flagellate *Giardia lamblia*. In the trophozoite stage, observe the size, shape, number of nuclei, and presence of flagella or pseudopodia. In the cyst stage, look for an increased number of nuclei and the thickened cyst wall.
   b. Sketch an example of each stage, label, and record in part 2 of the Laboratory Report.
3. Examination of protozoans present in stained blood slides:
   a. Examine with oil immersion a smear of blood infected with *Plasmodium vivax* and locate blood cells containing the parasite. After a mosquito bite, the parasites are carried to the liver, where they develop into **merozoites.** Later, they penetrate into the blood and invade the red blood cells, where they go through several stages of development. The stages are the delicate ring stage (see figure 20.7), the mature amoeboid form, and the **schizont** stage, in which the organism has divided into many individual infective segments that will then cause the red cell to rupture, releasing the parasites, which can then infect other cells. Sketch the red cells with the infective organism inside them, and note any changes in the red cell shape, pigmentation, or granules due to the effect of the parasite. Identify and label the stage or stages seen, and the species in part 3a of the Laboratory Report.
   b. Examine the trypanosome blood smear (see figure 20.2) with the oil immersion lens, and locate the slender flagellates between the red blood cells, noting the

flagellum and undulating membrane. Sketch a few red cells along with a flagellate in part 3b of the Laboratory Report.

4. Comparison of a free-living worm with its parasitic relative:

   a. Observe a prepared slide of a free-living flatworm (*Dugesia* species) with the low power objective. Note the pharynx, digestive system, sensory lobes in the head region, and the eyespots (see figure 20.3). Next examine a parasitic fluke such as *Clonorchis*. Note the internal structure, especially the reproductive system and eggs if female, and the organs of attachment such as hooklets or round suckers.

   b. Sketch each organism in part 4a of the Laboratory Report, and label the main features of each. Describe the main differences between the fluke and the free-living planaria.

   c. Examine prepared slides of a tapeworm (*Taenia* species, see figure 20.4), observing the small head, or scolex, and the attachment organs—the hooklets or suckers. Then locate along the worm's length the maturing proglottids. The smaller proglottids may show the sex organs better; a fully developed proglottid shows the enlarged uterus filled with eggs. Sketch, label, and describe its special adaptations to parasitic life in part 4b of the Laboratory Report.

5. Life cyle of *Schistosoma mansoni* and its importance in the control of schistosomiasis.

Assemble five or six slides showing the various stages in the schistosoma life cycle: adult worm (male and female if available), ova, ciliated miracidium, the sporocyst in the snail tissue, and the infective ciliate cercaria.

Next read this brief summary of the natural history of *Schistosoma mansoni* (see color plate 19 and **Nester et al. Microbiology: A human perspective 4th ed., 2004, pp. 317–319**).

Human schistosomiasis occurs wherever these conditions exist: water is polluted with human wastes; this water is used for human bathing and wading, or irrigation of cropland; and the presence of snail species that are necessary as hosts for the sporocyst stage in fluke development and completion of its life cycle. Solution to this public health problem is very complex, not only because of technical difficulties in its control and treatment, but also because its life cycle presents an ecological dilemma. Many developing countries need food desperately, but the main sources now available for these expanding needs are fertile deserts, which have adequate nutrients but require vast irrigation schemes, such as the Aswan Dam in Egypt. However, due to the unsanitary conditions and the presence of suitable snail hosts, these projects are accompanied by an increase in the disease schistosomiasis, which currently is very difficult to control and very expensive to treat on a wide scale.

The **cercaria** larvae swim in the contaminated water, penetrating the skin of agricultural workers who are barefoot. They migrate into the blood and collect in the veins leading to the liver. The *adults* develop there, copulate, and release the *eggs*. The eggs are finally deposited in the small veins of the large intestine, where their spines cause damage to host blood vessels. Some eggs die; however, others escape the blood vessels into the intestine and pass with the feces into soil and water. There they develop, and then hatch into motile **miracidia,** which eventually infect suitable snail hosts and develop into saclike **sporocysts** in the snail tissues. From this stage develop the fork-tailed cercaria larvae, which leave the snail and swim in the water until they die or find a suitable human host, thus completing the complex life cycle involving two hosts and five separate stages.

Now look at the prepared slides of all the schistosoma stages discussed in the description above. Sketch each stage in the appropriate place in the life cycle diagram shown in part 5 of the Laboratory Report.

EXERCISE

Laboratory Report: Parasitology:
Protozoa and Helminths

Results

1. Examination of Free-Living Cultures
   a. Pond water examination.
      Description of movements and drawings of any protozoans found in pond water.

   b. Examination of fresh samples of a free-living amoeba, paramecium, and flatworm. Description of movements and drawings with labels.

2. Examination of Stained Slides for Trophozoites and Cysts
   Prepare drawings of the trophozoite and cyst stage of either *Entamoeba histolytica* or *Giardia lamblia*. Label accordingly (see Procedure step 2).

3. Examination of Protozoans Present in Stained Blood Slides
   a. Examine blood smears of *Plasmodium vivax* (see Procedure step 3a).

   b. Blood smear of *Trypanosoma gambiense* (see Procedure step 3b).

4. Comparison of a Free-Living Worm with Its Parasitic Relative
   a. Comparison of *Dugesia* species (free-living) with *Clonorchis sinensis* (parasitic). See Procedure step 4a.

b. Study of a parasitic tapeworm (*Taenia* species). See Procedure step 4c.

5. Life Cycle of *Schistosoma mansoni* and Possible Methods of Control

   a. For each space in this life cycle, sketch the appropriate stage, using the prepared microscope slides.

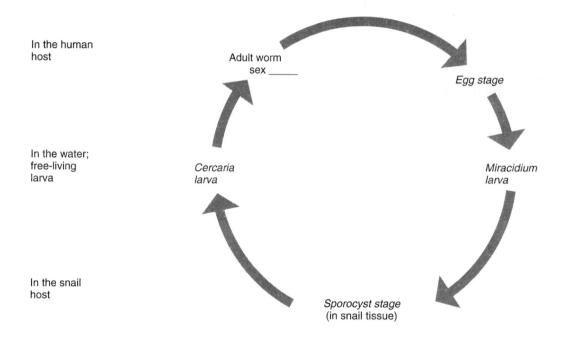

In the human
host

Adult worm
sex _____

Egg stage

In the water;
free-living
larva

*Cercaria
larva*

*Miracidium
larva*

In the snail
host

*Sporocyst stage
(in snail tissue)*

   b. Propose a plan for public health control of schistosomiasis. Describe various strategies that might be developed by public health personnel to interrupt this cycle and thus prevent schistosomiasis. Show on a diagram where specific measures might be taken, and label. Explain each possibility and its advantages and disadvantages.

Questions

1. Which form—the trophozoite or the cyst—is most infective when found in a feces sample? Explain.

2. In what ways are free-living and parasitic worms similar, such that they can be identified as closely related?

3. In what ways do the parasitic species differ from the free-living planaria? Use the chart to summarize your comparisons.

	Planaria	Fluke	Tapeworm
Outside covering			
Organs of attachment			
Sensory organs			
Digestive system			
Reproduction			

4. Estimate the length and width of a trypanosome. See figure 20.2 for a clue. Show your calculations.

5. How is the *Echinococcus* tapeworm transmitted to humans? Does it cause a serious disease? What are two ways in which its transmission to humans can be prevented?

NOTES:

Getting Started

Bacteriophage (usually shortened to phage) are viruses that infect bacteria. There are many kinds of bacteriophage, but this discussion is limited to DNA phage that have been well studied, such as lambda and T4. These phage first attach to the bacterial cell and inject their DNA into the cytoplasm. There are two major outcomes of this injection, depending on whether the phage are lytic or lysogenic.

1. **Lytic.** The cell lyses about 30 minutes after infection, releasing approximately 100 virus progeny (figure 21.1).

2. **Lysogenic** (or temperate). The DNA of the phage integrates into the bacterial chromosome and is replicated with the bacterial DNA. It may at some time in the future leave the chromosome, and direct the production of virus and lyse the cell. Bacteria that contain the DNA of a phage cannot be reinfected or lysed by the same type of phage.

Phage are too small (about 200 nm) to be seen in a light microscope, but can be detected if grown on a **bacterial lawn** as follows. Phage and their host cells are mixed in a small tube of soft agar and then

---

Figure 21.1 (*a-f*) Steps in the replication of a T-even phage during the infection of *E. coli*. From Eugene W. Nester et al. *Microbiology: A Human Perspective.* Copyright © 1998. The McGraw-Hill Companies. All Rights Reserved. Reprinted by permission.

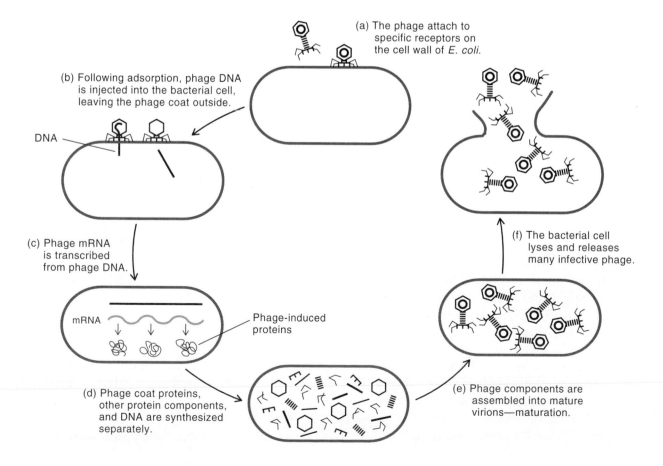

(a) The phage attach to specific receptors on the cell wall of *E. coli*.

(b) Following adsorption, phage DNA is injected into the bacterial cell, leaving the phage coat outside.

DNA

(c) Phage mRNA is transcribed from phage DNA.

mRNA

Phage-induced proteins

(d) Phage coat proteins, other protein components, and DNA are synthesized separately.

(e) Phage components are assembled into mature virions—maturation.

(f) The bacterial cell lyses and releases many infective phage.

Figure 21.2 Bacteriophage plaques formed on agar medium seeded with a lawn of bacteria. Courtesy of the University of Washington Photo Library.

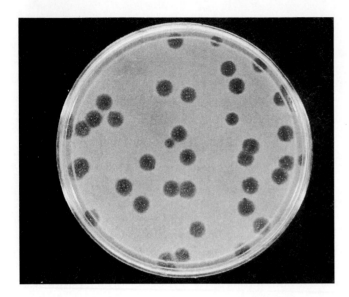

poured on top of an agar base plate. (Soft agar contains about half the concentration of standard agar so that the phage can diffuse more easily.) The plates are then incubated overnight at the optimum growth for the host bacteria.

During incubation, bacteria multiply and produce a thick covering of bacteria, or **bacterial lawn,** except in those places where phage have infected and killed the bacteria leaving clear areas called **plaques** (figure 21.2). Since each plaque originated with one phage, the plaques can be counted just as bacterial colonies to determine the number of phage originally mixed with the soft agar.

Although the appearance of the plaques can be influenced by many factors, in general **virulent phage** produce clear plaques. **Temperate phage** produce cloudy plaques because many cells within the plaque were lysogenized instead of lysed and thus continue to grow and multiply. The plaques do not increase in size indefinitely because phage can replicate only in multiplying bacteria.

Bdellovibrio, a small (1 × 0.25μm) bacterium that multiplies within bacterial cells, also forms plaques. Since it can grow in nondividing cells, its plaques continue to increase in size for a day or two. It is also found in sewage and can pass through the filters used to remove bacteria, and may be present in the filtrate you use for phage isolation.

It is important to study phage and to learn the techniques used to manipulate them for the following reasons.

1. Animal viruses, including human pathogens, are grown on tissue culture cells in the same fashion as phage on bacteria. Tissue culture are animal cells that are grown in bottles and plates. The animal virus can form plaques by causing cells to degenerate or die. Tissue culture cells require a more complex (and more expensive) medium, so it is convenient to learn viral technique with phage and bacteria.
2. Phage are used in recombinant DNA experiments and are also useful in studying the genetics of bacteria.
3. They are used to identify different strains of bacteria because one type of phage will only infect a few specific strains.
4. Lysogeny has served as a model for viruses inserting their DNA in animal cells. The life cycle of temperate phages resembles animal retroviruses.

In this exercise, you will attempt to isolate phage that infect *Escherichia coli* and learn to titer phage. Since sewage contains high numbers of *E. coli,* it is an excellent source of the *E. coli* phage. The sewage is filtered to remove bacteria but not the smaller viruses. Samples of the filtrate are then mixed with a laboratory strain of *Escherichia coli* in a suitable medium and observed for plaques on agar plates.

In the second part of the exercise, a suspension of phage is **serially diluted** so that an appropriate number of plaques can be counted on a plate and the titer of phage can be calculated. The number of phage/ml is the **titer.** The phage producing a plaque are also called plaque-forming units. Sometimes phage are present, but for some reason do not form a plaque.

Definitions

Bacterial lawn. The confluent growth of bacteria on an agar plate.

Bacteriophage. A virus that infects bacteria; often abbreviated phage (rhymes with rage).

Lysogen. A bacterium carrying a phage integrated in its chromosome. See temperate phage.

Lytic virus. A virus that replicates within a host cell and causes it to produce phage, rupture, and die. See virulent phage.

Plaque. A clear or cloudy area in a lawn of bacterial cells caused by phage infecting and lysing bacteria.

Plaque-forming units. The single phage that initiates the formation of a plaque.

Serial dilution. A dilution of a dilution continuing until the desired final concentration is reached.

Temperate, or lysogenic, phage. A phage that can either integrate into the host cell DNA or replicate outside the host chromosome, producing more phage and lysing the cell.

Titer. The concentration of virus in a sample (number/volume) when used as a noun, or to determine the concentration when used as a verb.

Virulent, or lytic, phage. A phage that always causes lysis of the cell following phage replication. Unlike a temperate phage, it cannot be integrated into the chromosome of the host.

Objectives

1. To isolate a phage from its natural habitat.
2. To titer a phage suspension.
3. To have an appreciation of the phage life cycle (to view steps in a phage life cycle).

References

Fraenkel-Convar, H. *Virology.* Englewood Cliffs, N.J.: Prentice-Hall, 1982.

Maniatis et al. *Molecular cloning,* 2nd ed. Volume I. New York: Cold Spring Harbor Laboratory Press, 1989.

Nester et al. *Microbiology: A human perspective,* 4th ed., 2004. Chapter 13.

Materials

First Session

Raw sewage filtered through 0.45 μm membrane filter

Host bacterial *Escherichia coli* K12 in a late log phase (OD~ 0.4)

Tubes of 4 ml soft tryptone overlay agar, 2

Tryptone agar base plates, 2

Part I: Bacteriophage Isolation and Culture from Filtered Sewage

Procedure

Safety Precautions: Filtered sewage may contain harmful animal virus. Handle with extreme care.

First Session

1. Label plates 1 ml and 0.1 ml (figure 21.3).
2. Melt overlay agar in boiling water and place in 50°C water bath for at least 5 minutes.
3. Add 1 ml filtered sewage to one tube soft overlay.
4. Add 0.1 ml filtered sewage to other tube of soft overlay.
5. Quickly add several drops (about 0.1 ml) of *E. coli* to each tube. Mix tubes and pour onto previously labeled tryptone agar base plates and rock gently to completely cover surface. If the tubes of overlay agar cool below 45°C, they will harden and the procedure must be repeated from the beginning.
6. Permit to harden for 5 minutes.
7. Incubate inverted at 37°C overnight.

Second Session

1. Examine the plates for evidence of plaque formation. Notice any different types of plaques and their relative sizes due to different kinds of phages. Count the number of each kind by placing dots with a marking pen under the plaques on the bottom of the petri plates. Record results.

Figure 21.3 Schematic showing dilution procedure for isolating bacteriophage from enriched sewage.

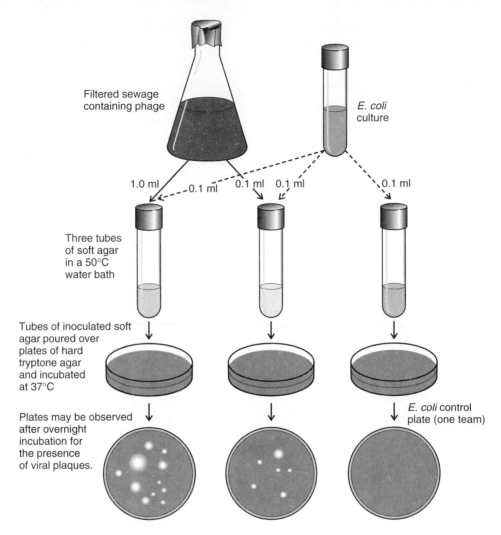

Filtered sewage containing phage

E. coli culture

1.0 ml — 0.1 ml — 0.1 ml — 0.1 ml — 0.1 ml

Three tubes of soft agar in a 50°C water bath

Tubes of inoculated soft agar poured over plates of hard tryptone agar and incubated at 37°C

Plates may be observed after overnight incubation for the presence of viral plaques.

E. coli control plate (one team)

2. Draw a circle the same size of each plaque on the bottom of the plate and reincubate. If any plaques are formed by *Bdellovibrio* the plaques will continue to enlarge. Phage plaques will remain the same size because, unlike *Bdellovibrio*, they can only reproduce in actively dividing cells.

Third Session

1. Examine plates for plaques that have increased in size.
2. If you do suspect *Bdellovibrio*, stab the plaque with a loop and prepare a wet mount. Look for very small, motile bacteria.
3. Record results.

Part II: Titering a Phage Suspension

> **Materials**
>
> Host bacteria for phage in late log phase
> Per team
>   *Escherichia coli*
>   phage T4 suspension
>   9-ml tryptone blanks, 4
>   4-ml overlay agar tubes, 4
>   Tryptone agar base plates, 4
>   Sterile 1-ml pipets, 5

Procedure

First Session

1. Label four 9-ml tryptone blanks: 10^{-1}, 10^{-2}, 10^{-3}, 10^{-4} (figure 21.4).

2. Transfer 1 ml of the bacteriophage to the tube labeled 10^{-1} with a sterile 1-ml pipet. Discard the pipet. You must use fresh pipets each time so that you do not carry over any of the more concentrated phage to the next dilution.

Figure 21.4 Schematic showing the procedure used to demonstrate *E. coli* phage plaques on the surface of agar plates. Plaques are represented by the light-colored areas on the 10^{-3} and 10^{-4} dilutions. (These results are only one possibility.)

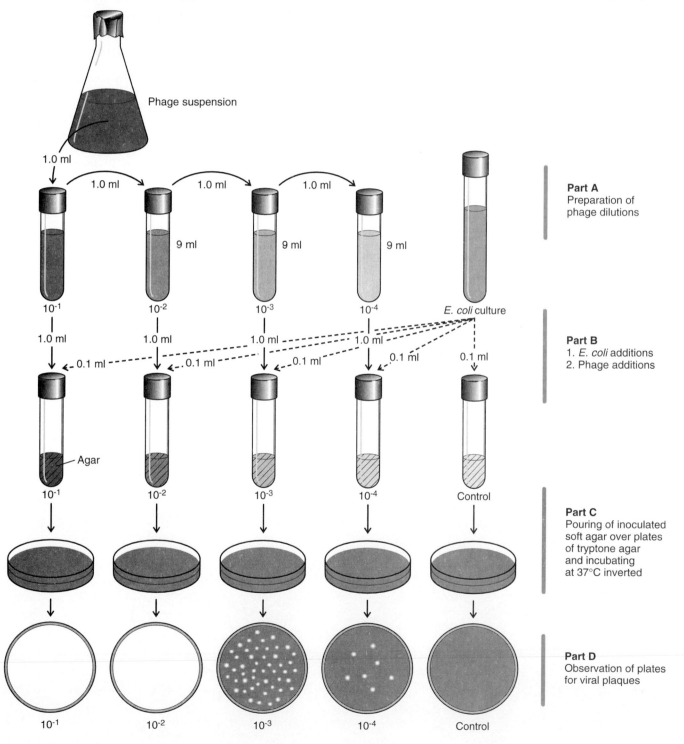

3. Mix and transfer 1 ml of the 10^{-1} dilution to the 10^{-2} tube and discard pipet.
4. Mix and transfer 1 ml to the 10^{-3} tube and discard pipet.
5. Mix and transfer 1 ml to the 10^{-4} tube and discard the pipet.
6. Label four tryptone hard agar petri plates: 10^{-1}, 10^{-2}, 10^{-3}, 10^{-4}.
7. Melt four tubes of soft overlay agar and place in a 50°C water bath. Let cool for about 10 minutes.
8. Add about 0.1 ml (or several drops) of *E. coli* broth to each tube of melted overlay agar.
9. Starting with the most diluted phage tube (10^{-4}), add 1 ml to the overlay agar and immediately pour on the tryptone agar base plate labeled 10^{-4}.
10. Using the same pipet, add 1 ml of the 10^{-3} dilution to a tube of overlay agar and pour into the plate labeled 10^{-3}. You can use the same pipet because you started with the most dilute sample and therefore the phage carried over are minimal.
11. Repeat for the 10^{-2} and 10^{-1} phage dilution.
12. Incubate the plates inverted at 37°C after the agar has hardened.

Second Session

1. Examine the plates. Select a plate containing between 30 and 300 plaques. As you count the plaques, place a dot with a marking pen under each plaque on the bottom of the petri plates. These marks can be wiped off so that each team member can count the plaques.
2. Estimate the numbers on the other plates. They should vary by a factor of 10 as the dilution increases or decreases.
3. To determine the titer, use this formula:

No. of plaques × 1/dilution × 1/ml of sample = plaque forming units/ml.

Example: If 76 plaques were counted on the 10^{-4} dilution, then:

$$76 \times 1/10^{-4} \times 1/1 = 76 \times 10^4 \text{ pfu/ml.}$$

4. Record results.

EXERCISE

Laboratory Report: Prokaryotic Viruses:
Bacteriophage Isolation and Titering

Results

1. Isolation and Culture from Filtered Sewage

    a. How many different types of plaques observed? _____

        type 1 appearance _____ number _____

        type 2 appearance _____ number _____

        type 3 appearance _____ number _____

    b. Did any plaques increase in size after reincubating? _____

        If yes, were small, very motile bacteria from the edge of the plaque observed?

2. Titering a Phage Suspension

Dilution numbers of plaques	Control	10^{-1}	10^{-2}	10^{-3}	10^{-4}

    a. Which dilution resulted in a countable plate?

    b. Did the number of plaques decrease 10-fold with each dilution?

    c. How many phage/ml were in the original suspension? Show calculations (see exercise 8).

Questions

1. Why was the sewage sample filtered?

2. How can you distinguish a lytic phage from a temperate phage when observing plaques from the filtered sewage sample?

3. Why can a plaque be considered similar to a bacterial colony?

4. Why do plaques formed by *Bdellovibrio* continue to increase in size after 24 hours, but not plaques formed by phage?

INTRODUCTION to Medical Microbiology

It is easy to think of microorganisms as a deadly, vicious force—especially when the diseases they cause kill young people or wipe out whole populations. The organisms, however, are simply growing in an environment they find favorable.

If pathogens become too efficient at taking advantage of their host, the host dies and the organism dies with it. Thus, the most successful pathogens are those that live in balance with their host. When a new pathogen enters the population, it is very virulent, but after awhile there is a selection toward less virulent pathogens and also a selection in the hosts for increased resistance.

Medical microbiology continues to offer challenges to those interested in medicine and in pathogenic bacteria. These next exercises are an introduction to many of these organisms that are encountered in a clinical laboratory. Not only will you study the characteristics of the organisms, but also you will learn some strategies for isolating and identifying them. In addition, these exercises are designed to help you learn to differentiate between organisms you can expect to find as normal flora in various places in the body and others that are responsible for certain diseases.

NOTES:

EXERCISE

Normal Skin Flora

Getting Started

The organisms growing on the surfaces and in the orifices of the body are called normal flora. They are usually considered **commensals** because they do not harm their host, and, in fact, have several beneficial roles. Normal flora prevent harmful organisms from colonizing the skin because they are already established there and utilize the available nutrients. Some produce enzymes or other substances that inhibit nonresident organisms. Other organisms, called transients, can also be found on the skin for short periods, but they cannot grow there and soon disappear.

Familiarity with organisms making up the skin flora is useful because these organisms are frequently seen as contaminants. Skin is continually flaking off, and bacteria floating in the air on rafts of skin cells sometimes settle into open petri dishes. If you are familiar with the appearance of *Staphylococcus* and *Micrococcus* colonies, for instance, you will be able to suspect contamination if you see such colonies on an agar plate. *Staphylococcus epidermidis* can also be seen in clinical specimens such as urine samples. These organisms probably are not causing disease, but are simply contaminants from skin flora.

Some of the organisms you may isolate:

Staphylococcus epidermidis This Gram-positive coccus is found on the skin as part of the normal flora of almost all humans throughout the world. It can also be isolated from many animals.

Staphylococcus aureus At least 20% of the population "carry" (have as part of their normal flora) this bacterium. It is found on the skin, especially in the nares or nostrils, and it seems to cause no harm to its host. However, *S. aureus* is frequently the cause of wound infections and food poisoning, and has been implicated as the cause of toxic shock syndrome. It can be identified by the **coagulase test.** Recently, many additional species of *Staphylococcus* have been identified. They are associated mostly with diseases in immunologically compromised individuals.

Micrococcus luteus This is a Gram-positive coccus found on the skin of some people, but it almost never causes disease. It is frequently an air contaminant forming bright yellow colonies.

Propionibacterium acnes These anaerobic, Gram-positive rods are **diphtheroid** or club shaped. When investigators tried to isolate an organism that might be the cause of acne, they almost always found the same Gram-positive diphtheroid rods in the lesions. Therefore, they named the organism *Propionibacterium acnes*. However, when people without acne were studied, it was found that *P. acnes* was present on their foreheads as well. Although some people have a much higher population of these organisms than others, the number of organisms does not seem to correlate with acne or any other skin condition.

Propionibacterium granulosum A Gram-positive diphtheroid rod found on some individuals, usually in smaller numbers than *P. acnes* (figure 22.1). It is considered a harmless commensal.

Figure 22.1 Colonial appearance of some normal skin flora organisms in a Gram stain.

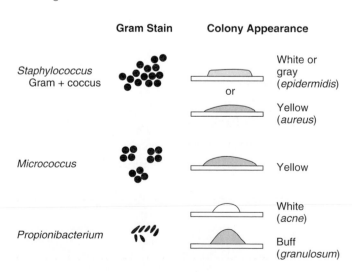

Definitions

Coagulase test. A test in which organisms are mixed with plasma on a slide. If the cells clump together, the culture is coagulase positive.

Commensals. Organisms that live together in close association and may or may not benefit each other.

Diphtheroid. A Gram-positive, club-shaped organism sometimes called a coryneform. *Propionibacterium* and *Corynebacterium* are examples of diphtheroid-shaped organisms.

Objectives

1. To learn to identify organisms making up the normal skin flora.
2. To understand the importance of skin flora.
3. To learn about the anaerobe jar.

References

Marples, Mary J. "Life on the human skin." *Scientific American*, January 1969, 220(1):108–115.

Nester et al. *Microbiology: A human perspective*, 4th ed., 2004. Chapter 22, Section 22.2.

Materials

First Session

  TSY agar plates (or TSY contact plates 2 inches in diameter), 2

  GasPak anaerobe jar (or other anaerobic system)

  Sterile swab

  Sterile saline

  70% ethanol

  Cotton balls

Second Session

  Tubes of sterile water (0.5 ml/tube), 6

  TSY + glucose agar deeps (yeast extract and glucose are added to TS agar to encourage the growth of *Propionibacterium*), 4

  TSY + Glucose + bromcresol purple agar slants, 2

  Magnifying glass is optional but helpful

Third Session

  Plasma

Procedure

Safety Precautions: Some students may isolate *Staphylococcus aureus* as part of their normal flora. This is a pathogen and should be handled with extra care.

First Session

1. Saturate a cotton ball with 70% ethanol and rub the forehead for 20 seconds. This will remove any transient organisms you might have on your skin, which are not part of your normal flora.
2. Let the forehead dry for about 20 minutes. Avoid touching it with your hair or fingers.
3. Moisten a sterile swab with saline and rub it briskly on an area of your forehead about the size of a quarter for about 15 seconds.
4. Immediately swab the first third of a TSY agar plate, discard the swab, and finish the streak plate with a loop.
5. Repeat the procedure, swabbing a second TSY agar plate from an adjacent area of the forehead.

   (Alternative method to steps 3, 4, and 5: Press an open contact TSY agar plate on the forehead. Repeat with a second plate on an adjacent area.)
6. Incubate one of the TSY agar plates aerobically at 37°C. Incubate the second TSY agar plate anaerobically in a GasPak or other anaerobe jar at 37°C. Follow the manufacturer's directions for creating an anaerobic atmosphere.

7. After 48 hours of incubation, the student or instructor should store the aerobic plate at room temperature to prevent the plate from drying out. *Staphylococcus* and *Micrococcus* can be observed after 48 hours, but *Propionibacterium* must be incubated five days before colonies can be seen.

Second Session
(5 days later)

1. Examine the aerobic TSY plate and circle two different colony types with a marking pen on the bottom of the plate. Make a Gram stain of each circled colony.
2. If the colonies are Gram-positive cocci, suspend the remainder of the colony used for the Gram stain in 0.5 ml sterile saline. Use this suspension to inoculate:
   a. a glucose + bromcresol purple TS agar slant
   b. a cooled melted agar deep (sometimes called a shake tube). See exercise 9 for inoculation procedure. Incubate at 37°C.
3. Examine the plate incubated in the anaerobe jar. You will see some of the same colony types observed on the aerobic plate, because *Staphylococcus* are facultative anaerobes and can grow with or without oxygen. *Propionibacterium* colonies, however, are white and very small—only a few millimeters in diameter. *P. granulosum* are slightly larger and appear as buff or pink cones when the plate is examined with a magnifying glass. Choose two possible *Propionibacterium* colonies, Gram stain them, and if they are diphtheroid Gram-positive rods, inoculate into a shake tube. You can inoculate the shake tube directly from the plate if very little remains of the colony.

Third Session
(5 days later)

1. Observe the glucose + bromcresol slants. If the organism is able to ferment glucose, the acid produced will turn the purple agar yellow.
2. Observe the agar deeps. Obligate aerobes are only able to grow on the top, while facultative anaerobes will grow throughout the entire tube. The obligate anaerobes will not be able

to grow in the top few centimeters where oxygen has diffused in—only in the bottom anaerobic portion. (See exercise 9.)
3. Identify your isolates. The following is a description of the organisms most commonly found on the forehead.

Staphylococcus are Gram-positive cocci arranged in clusters. They are facultative and can ferment glucose. There are two main species of *Staphylococcus* found on the skin: *epidermidis* and *aureus*. *S. aureus* tends to have yellow colonies and *S. epidermidis* white colonies.

Optional: If you have yellow colonies of *Staphylococcus*, you can determine if it is *S. aureus* with a coagulase test. *S. aureus* is coagulase positive and *S. epidermidis* is coagulase negative.

Micrococcus are Gram-positive cocci arranged in packets of four or eight. They are obligate aerobes and cannot ferment glucose or other sugars.

Propionibacterium acnes are diphtheroid coryneform Gram-positive rods that tend to palisade or line up like a picket fence. They form very small, white, glistening colonies. *Propionibacterium* are obligate anaerobes that grow only in the absence of oxygen, but are not killed by air as are some anaerobes.

Propionibacterium granulosum appears the same as *P. acnes* in a Gram stain but forms pinkish, slightly larger colonies.

Coagulase Test

Place a drop of water on a slide and make a *very* thick suspension of cells from a yellow colony.

Place a drop of plasma next to it and mix the two drops together. Look for clumping; clumped cells indicate a coagulase-positive result.

Drop the slide in boiling water and boil for a few minutes to kill the organisms before cleaning the slide.

	Summary of Reactions			
	Gram stain	**Colony Color**	**Metabolism**	Glucose
Staphylococcus epidermidis	+ cocci	white/gray	facultative	+
Staphylococcus aureus	+ cocci	yellow	facultative	+
Micrococcus	+ cocci	yellow	obligate aerobe	–
Propionibacterium	+ rods	white/buff	obligate anaerobe	

EXERCISE

Laboratory Report: Normal Skin Flora

Results

	Isolate 1	Isolate 2	Isolate 3	Isolate 4
Gram stain				
Colony appearance				
Glucose fermentation				
Agar deep				
Possible identity				

* Although this was not a quantitative procedure, what organism seemed to be the most numerous on your forehead?

Questions

1. How could normal skin flora be helpful to the host?

2. Why did you wipe your forehead with ethanol before sampling it?

3. Why was *Staphylococcus* the only organism that could grow on both plates?

4. How can you immediately distinguish *Staphylococcus* from *Propionibacterium* in a Gram stain?

5. Why does *Staphylococcus* probably cause more contamination than *Propionibacterium*, even though most people have higher numbers of the latter? (Hint: Are most agar plates incubated aerobically or anaerobically?)

EXERCISE

Respiratory Microorganisms

Getting Started

In this exercise, you have an opportunity to observe a throat culture and learn some of the ways a clinical microbiologist identifies pathogenic organisms. You also examine the normal flora of the throat, and while you will not work with actual pathogens, many of the organisms in the normal flora resemble related pathogenic bacteria. In exercises 29 and 30, you use **serological techniques** to further differentiate some of these organisms.

A physician frequently orders a throat culture if a patient has a very sore throat and fever. This is done to determine whether or not the sore throat is caused by Group A β-hemolytic *Streptococcus pyogenes*. This organism is important because it not only causes strep throat, a disease characterized by fever and a very sore throat, but also two very serious diseases can follow the original infection: rheumatic fever, a disease of the heart, and acute glomerulonephritis, a disease of the kidney.

Fortunately, streptococci are usually still sensitive to penicillin and related antibiotics, and treatment is fairly straightforward. However, most sore throats are caused by viruses. Since viruses do not have a cell wall or any metabolic machinery, they are not affected by penicillin or most other antibiotics. Therefore, it is important to make an accurate diagnosis so that antibiotics can be used wisely.

Other pathogens sometimes found in the throat are *Streptococcus pneumoniae*, *Neisseria meningitidis*, *Haemophilus influenzae*, and *Klebsiella pneumoniae*. If these organisms are indeed causing disease, they are usually present in large numbers and the patient has the symptoms of the disease.

β-hemolytic *Streptococcus* can be identified when growing on blood agar. This medium is made of a base agar that is rich in vitamins and nutrients. Before pouring the melted agar into the petri plates, 5% sheep blood is added. (Sheep raised for this purpose donate blood as needed.) The blood performs two functions: it adds additional nutrients and aids in distinguishing α-hemolytic from β-hemolytic streptococci.

Streptococci produce hemolysins that act on red blood cells (also called erythrocytes). α-hemolytic streptococci, which are a major component of the normal throat flora, incompletely lyse red blood cells. When the organism is growing on blood agar, a zone of partial clearing of the red blood cells can be seen around a colony. β-hemolytic streptococci produce hemolysins that completely lyse the red blood cells and therefore produce a clear zone in the blood agar around the colony. It is important to note that β-hemolysis is not always correlated with pathogenicity. For example, some strains of *E. coli* can produce β-hemolysis but are not responsible for any disease.

Commercial kits based on specific antibodies or other methods are now available that can be used to determine if a patient has strep throat. Although these tests can be performed in the doctor's office, the results sometimes must be verified by growing the culture on blood agar.

A small percentage of the hemolysins of β-hemolytic streptococci are oxygen labile, which means they are destroyed by oxygen. In a clinical laboratory, throat cultures are incubated in an anaerobe jar so that no hemolysis is overlooked. In this laboratory exercise, we will incubate the cultures in a candle jar, if convenient. This increases the CO_2 content of the atmosphere and enhances the growth of some organisms, but does not create anaerobic conditions.

The throat contains a plethora of organisms that make up the normal flora, many of which resemble pathogenic organisms. Frequently, some actual pathogenic organisms are found in small numbers, such as *Streptococcus pneumoniae* or β-hemolytic *Streptococcus*. The presence of these organisms is only significant when they appear in large numbers.

Figure 23.1 Normal flora of the throat.

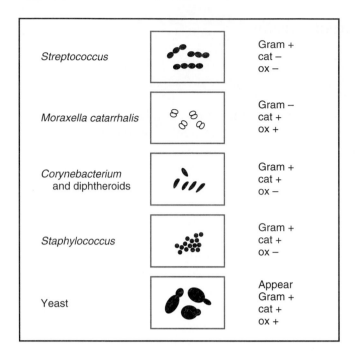

Streptococcus		Gram + cat – ox –
Moraxella catarrhalis		Gram – cat + ox +
Corynebacterium and diphtheroids		Gram + cat + ox –
Staphylococcus		Gram + cat + ox –
Yeast		Appear Gram + cat + ox +

Examples of common normal flora of the throat are diagrammed in figure 23.1. You might observe some of them on your streak plate.

α-hemolytic streptococcus These organisms will be the predominant organisms growing on your plates. Another name for them is Viridans streptococci, and they rarely cause disease. They are Gram-positive and grow in short chains. They are catalase negative, which distinguishes them from *Staphylococcus*. *Streptococcus pneumoniae* is also *α*-hemolytic and is differentiated from the normal flora by sensitivity to certain antibiotics and serological tests. They are all **oxidase** negative.

Moraxella catarrhalis These organisms are Gram-negative cocci arranged in pairs. They resemble the microscopic morphology of *Neisseria* (which grows only on a different kind of heated blood agar called chocolate agar). The genus *Neisseria* includes such pathogens as *N. gonorrhoeae* and *N. meningitis*. Try to find a colony of *Moraxella* and Gram stain it so you can be familiar with the appearance of these organisms. The colonies may be a little larger than *Streptococcus* and are oxidase positive, so if oxidase reagent is available, it will help you identify a colony. They are also **catalase** positive.

Note: *Moraxella catarrhalis* was formerly named *Branhamella catarrhalis*, and prior to that *Neisseria catarrhalis*.

Corynebacterium and Diphtheroids These are irregular, club-shaped Gram-positive rods. They are part of the normal flora but resemble *Corynebacterium diphtheriae*, which causes diphtheria. They are usually catalase positive and oxidase negative.

Staphylococcus These Gram-positive cocci are arranged in clusters. *Staphylococcus aureus* frequently is part of the normal flora, although it is a potential pathogen. The colonies are usually yellow. *Staphylococcus* is catalase positive and oxidase negative.

Yeasts These are fairly common in the oral flora and they form relatively large colonies. In a Gram stain, the cells (which are eukaryotic) appear purple, are larger than bacteria, and sometimes have buds. They are catalase positive and usually oxidase positive.

Definitions

Catalase.   An enzyme found in most aerobic organisms that breaks down H_2O_2 to water and oxygen.

Oxidase.   A reagent that tests for cytochrome C.

Serological test.   Identification of organisms by mixing cells with serum containing antibodies to a specific organism. If the cells clump, the test is positive for that organism.

Objectives

1. To learn the importance of Group A *β*-hemolytic *Streptococcus* and how to distinguish it from normal flora.
2. To observe normal flora of the throat.

References

Holt, John G. et al. *Bergey's manual of determinative bacteriology*. Baltimore: Williams & Wilkins, 1994.

Nester et al. *Microbiology: A human perspective*, 4th ed., 2004. Chapter 23.

Ryan, Kenneth J., ed. *Sherris medical microbiology, an introduction to infectious diseases*, 3rd ed. Norwalk: Appelton and Lange, 1994.

Materials

Per student

First Session

Blood agar plate, 1

Sterile swab, 1

Tube of sterile saline, 1

Tongue depressor

Demonstration cultures

α-hemolytic *Streptococcus* on blood agar

β-hemolytic *Streptococcus* on blood agar

Candle jar

Second Session

H_2O_2

Oxidase reagent

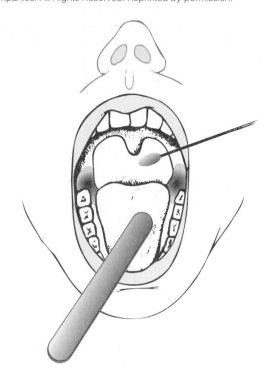

Figure 23.2 Diagram of open mouth. Shaded areas indicate places to swab. From Josephine A. Morello, Helen Eckel Mizer, and Marion E. Wilson, *Laboratory Manual and Workbook in Microbiology: Applications to Patient Care.* Copyright © 1994 The McGraw-Hill Companies. All Rights Reserved. Reprinted by permission.

Procedure

Safety Precautions: There may be colonies of β-hemolytic streptococci (*Streptococcus pyogenes*) and *Staphylococcus aureus* on the agar plates of normal flora. Handle these plates and the demonstration plates of β-hemolytic *streptococcus* with special care.

First Session

1. Swab your partner's throat. First seat your partner on a stool and shine a light on the throat. Carefully remove a sterile swab from the wrapper and moisten with the sterile saline, pressing out excess moisture on the inside of the tube. Depress the tongue with the tongue depressor and swab the tonsilar area on the side of the throat (figure 23.2). Do not swab the hard palate directly in the back behind the uvula and do not touch the tongue or lips. Do this rather quickly to avoid the gag response.

2. Swab the first third of a streak plate on the blood agar plate, rolling the swab over the agar to be sure to inoculate all sides. Discard the swab and continue streaking the rest of the plate with your loop as usual for isolated colonies.

3. Incubate your plate at 37°C in a candle jar, if available.

Second Session

1. Examine your throat culture plate and compare it to the pure cultures of α- and β-hemolytic streptococci on the demonstration plates. If you observe β-hemolysis, make a Gram stain of the colony to determine if it is a Gram-positive streptococcus. If they indeed appear to be β-hemolytic streptococcus colonies, are they the predominant colony type on the plate? In a clinical laboratory:

   a. If they were the most numerous colony type, the physician would be notified and the patient would be treated.

   b. If only a few colonies were present, then the results could be reported as "ruled out β strep."

2. Make Gram stains of various colonies and do catalase and oxidase tests on the same colonies if possible. Record the kinds of organisms you observed based on their Gram stain, colony morphology, catalase test, and oxidase test. See Getting Started for a description of some of the organisms you might see.

Tests

Catalase. With a sterile loop, place some cells from the colony to be tested on a glass slide. Cover the cells with a few drops of H_2O_2. If bubbles are formed, the culture is catalase positive (figure 23.3). Boil the slide in water for a few minutes to kill the organisms.

Oxidase. Place a small piece of filter paper on a glass slide and moisten with freshly prepared oxidase reagent. Remove some cells from a colony to be tested with a sterile loop and rub the loop on the moistened filter paper. If a pinkish-purple color appears, the cells are oxidase positive (see figure 23.3). Place the paper in an autoclave bag and boil the slide for a few minutes to kill the organisms.

Figure 23.3 (*a*) Catalase and (*b*) oxidase tests.

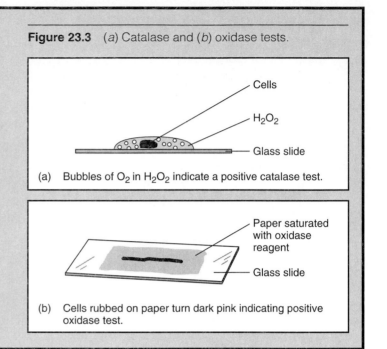

(a)   Bubbles of O_2 in H_2O_2 indicate a positive catalase test.

(b)   Cells rubbed on paper turn dark pink indicating positive oxidase test.

EXERCISE

Laboratory Report: Respiratory Microorganisms

Results

Indicate the numbers of organisms, using +++ for the most numerous or highest number, and + if very few are observed. Use ++ for numbers in between.

	Appearance of Hemolysis	Numbers Present
α-hemolytic streptococci		
β-hemolytic streptococci		

Other organisms observed:

Possible Identity	Gram Stain	Catalase	Oxidase	Numbers Present

Questions

1. What is the predominant organism in your throat flora? Did you observe any β-hemolytic streptococci in your throat culture?

2. What is the difference between alpha- and beta-hemolysis?
   a. On the red blood cells?

   b. On the blood agar plate?

3. Give two reasons it is very important to correctly diagnose and treat strep throat.

4. Name one genus of Gram-negative cocci.

5. If a student had a cold and sore throat caused by a virus, how would the virus appear on the blood agar plate?

EXERCISE

24

Identification of Enteric Gram-Negative Rods

Getting Started

In this exercise, you will learn how to identify bacteria with biochemical tests. Since it is frequently necessary to identify Gram-negative rods in the clinical laboratory, these are excellent organisms to use as examples of the general procedure. **Enteric** organisms, for instance, are part of the normal flora of the intestine, but are responsible for the majority of the urinary tract infections. At times, organisms from the colon (large intestine) can infect the bladder and grow in the urine stored there.

Biochemical tests measure such things as the differences in the organisms' ability to ferment different sugars, the presence of various enzymes, and physical characteristics such as motility. An organism then can be identified by comparing the results of the tests to the results of known bacteria in such reference books as *Bergey's Manual of Systematic Bacteriology*. The following is a discussion of the tests you will be using and how they work.

Fermentation Tubes Carbohydrate fermentation tubes consist of a complete broth, a carbohydrate such as the sugar glucose, and the pH indicator andrades. A small Durham tube is added, which fits inside the large tube. If the organism can ferment the sugar, it will produce acidic products and the pH will fall, turning the pH indicator pink. If the organism also produces a gas such as hydrogen, some of it will accumulate in the Durham tube as a bubble (see color plate 20).

After incubation, you should examine the tubes for growth (sometimes you have to shake the tubes slightly because the cells have settled). If for some reason there is no growth, the test must be repeated. If there is growth, the results can be recorded as:

A   Acid production—if the indicator has turned red
A/G   Acid and Gas—if the indicator has turned red and a bubble is seen in the Durham tube

N/C   No Change—if neither gas nor acid has formed

The sugars glucose, sucrose, and lactose are especially useful in the identification of the enteric Gram-negative rods. Lactose fermentation aids in the preliminary differentiation of enteric pathogens from the normal **coliforms.** *Salmonella* and *Shigella* species, which cause enteric diseases of various kinds, do not ferment lactose; however, members of the common fecal flora, *Escherichia coli*, *Enterobacter aerogenes*, and *Klebsiella* are able to ferment lactose. Lactose fermentation is not related to pathogenicity in any way, it is simply a convenient characteristic for identifying organisms. The enteric *Proteus*, for example, is lactose negative, but usually a nonpathogen.

Methyl Red and Voges-Proskauer The fermentation tubes previously described show whether or not fermentation has taken place. These two tests identify a particular kind of fermentation.

Fermentation is the energy-yielding pathway a facultative organism can use if oxygen is unavailable. Organisms have different fermentation pathways resulting in different end products. Some are simple, such as the conversion of pyruvate to lactic acid. Other pathways are more complex, yielding a variety of products and perhaps additional energy (figure 24.1).

Methyl Red This test measures the pH after organisms have grown in buffered peptone glucose broth (MR/VP broth). It is positive when the pH is less than 4.3. *E. coli* and other organisms ferment sugars by the mixed acid pathway. The products of this pathway are predominantly acetic and lactic acids, some organic compounds, and CO_2 and H_2. This results in a very low pH and therefore *E. coli* is methyl red positive.

Other bacteria such as *Enterobacter* use the butanediol pathway when fermenting sugars. The end products are predominantly alcohols and a small

Figure 24.1 Fermentation pathways.

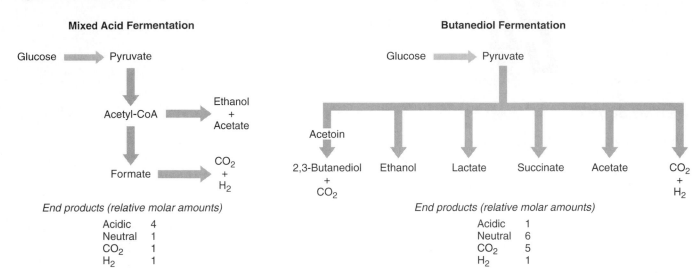

amount of acid, as well as CO_2 and H_2. Although the pH is low, it is not low enough to change the methyl red, so these bacteria are methyl red negative.

Voges-Proskauer One of the intermediates in the butanediol pathway is acetoin, for which the Voges-Proskauer reagents test. If the test is positive, then acetoin is present and the organism is using the butanediol pathway. The reagents Voges-Proskauer A and B are added to the broth culture after incubation. If a brick red precipitate forms, it is VP+.

Indole Some organisms have an enzyme that cleaves the amino acid tryptophan, producing indole. They can be grown in tryptone broth, which has a high level of tryptophan. After incubation, the broth is tested for the presence of indole by adding Kovacs reagent. A red ring forms on the top of the broth if the organism is indole+. See color plate 21.

Citrate Utilization The organism is grown on Simmons citrate medium, a mineral medium with citrate as the sole carbon source. The agar also contains the pH indicator bromthymol blue. If the organism can utilize the citrate, the pH rises and the indicator turns a deep blue. See color plate 22.

Urea Hydrolysis Organisms are grown on agar containing urea and a pH indicator. If the organism produces the enzyme urease, urea is split, forming ammonia and CO_2. This raises the pH of the medium, turning it bright pink. *Proteus* species can be distinguished from other enteric rods because it always produces urease. See color plate 23.

Motility Organisms are stabbed into a semisolid agar deep containing tetrazolium, an electron acceptor that turns red wherever there is growth. After incubation, a diffuse red color throughout the medium indicates motility. If there is a red streak only where the agar was stabbed, then the organism is nonmotile.

MacConkey Agar Only Gram-negative organisms grow on this medium. If the colonies are purple, the organism ferments lactose.

In this exercise, each team inoculates a series of biochemical tests with a labeled organism and an unlabeled "unknown" organism. (The "unknown" is one of the organisms listed.) There may seem to be a large number of tubes to inoculate, but if they are labeled and lined up in a test tube rack, inoculation can be done fairly quickly. Good organization is very helpful.

Note: Commercial test systems (such as *Enterotube II System*) are available in which a whole series of tests are inoculated at the same time (see color plate 24). After incubation, the results are read and the organism can be identified.

Definitions

Coliforms. Gram-negative rods found in the intestine that ferment lactose with the production of acid and gas—*Escherichia coli*, *Enterobacter*, and *Klebsiella*.

Enteric. Associated with the intestine.

Fermentation. An energy-yielding metabolic pathway in which organic compounds are both electron donors and acceptors.

Objectives

1. To learn to identify bacteria using biochemical tests.
2. To understand the physiological basis for the tests.
3. To become familiar with organisms commonly seen in a clinical laboratory, especially those causing urinary tract infections.

Reference

Nester et al. *Microbiology: A human perspective,* 4th ed., 2004. Chapter 10, Section 10.2.

Materials

Per team of two students

Cultures on TS agar slants of

  Escherichia coli

  Enterobacter aerogenes

  Klebsiella pneumoniae

  Pseudomonas aeruginosa

  Proteus mirabilis

Sterile saline 1.0 ml/tube, 2 tubes

Glucose fermentation tubes, 2

Lactose fermentation tubes, 2

Sucrose fermentation tubes, 2

Tryptone broths, 2

Methyl red-Voges-Proskauer (MR-VP) broth tubes (one tube for each test), 4

Simmons citrate slants, 2

Motility agar deeps, 2

MacConkey agar plates, 2

Urea slants, 2

Reagents

  Oxidase reagent (freshly prepared) and small squares of white filter paper

  Kovacs reagent

  Methyl red

  Voges-Proskauer reagents A and B

Procedure

First Session

1. Choose one labeled culture and one "unknown" culture (or the cultures you are assigned).
2. Label a set of tubes for each organism with the name of the organism, the medium, the date, and your name or initials. Note that you will need 1 MR-VP broth for the methyl red test and another MR-VP broth for the Voges-Proskauer test. You should have a total of ten tubes for each organism to be identified.
3. Make a suspension of each organism by adding bacteria with a sterile loop to the sterile saline. Use this suspension to inoculate the test media.
4. Inoculate the tubes by adding a loopful of the suspended organisms to the broth. Inoculate the agar slant by gliding the loop over the surface of the agar, starting at the bottom of the slant. Streak the MacConkey plates with a loop full of suspension to obtain isolated colonies. Use good aseptic technique to avoid contamination with unwanted organisms.
5. Inoculate the motility deep by stabbing the agar almost to the bottom with your inoculating loop. Use cells from the original slant instead of from the suspension to be sure you have enough cells.
6. Perform an oxidase test using cells from the slant. See exercise 23.
7. Incubate the tubes at 37°C for 2 days or more. Incubate the motility deeps at room temperature. Some organisms are not motile at 37°C.

Second Session

1. Examine the fermentation tubes and record as A if the pH indicator has turned red from acid production, G for gas production, and N/C for no change if neither acid nor gas has formed. Be sure there is growth in the tube before recording (see plate 20).
2. Add a dropper full of Kovacs reagent to the tryptone broth and shake slightly. A red layer on the top of the broth indicates a positive test for indole (see plate 21).

3. Add a few drops of methyl red to one of the MR-VP broths. A red color indicates a positive test.
4. Add 10–15 drops of V-P reagent A (alpha-naphthol solution) and 10–15 drops of V-P reagent B (40% KOH). Shake and let stand a few minutes, or an hour maximum. The appearance of a red color indicates a positive test.

   Safety Precaution: Alpha-naphthol is toxic.
5. Observe citrate slant. A deep blue color is positive for citrate utilization (see plate 22).
6. Examine the motility deep. If the tube appears pink throughout the agar, it is positive for motility. If only the original stab line appears pink, the test is negative for motility.

7. Observe colonies on MacConkey plate. Purple colonies indicate lactose fermentation.
8. Observe urea slant. A bright pink color is positive for urea hydrolysis (see plate 23).
9. Record the results of your known organism in the chart below and on a similar chart drawn on the blackboard. These will be the results to establish the reactions of the known bacteria. Consult your instructor if student results of their known bacteria do not agree.
10. Record the class results on the board. Also record the results of your unknown organism.

| | Fermentation | | | | | | | | | | |
	glu	lac	suc	indole	MR	VP	cit	mot	ox	mac	urea
E. coli											
Enterobacter											
Klebsiella											
P. mirabilis											
Pseudomonas											

EXERCISE

Laboratory Report:
Identification of Enteric Gram-Negative Rods

Results

	Fermentation			indole	MR	VP	cit	mot	ox	mac	urea
	glu	lac	suc								
E. coli											
Enterobacter											
Klebsiella											
P. mirabilis											
Pseudomonas											
Unknown											

Questions

1. What is the identity of your unknown organism?

2. Can you determine whether an organism can ferment a sugar if it does not grow in the broth? Explain.

3. How can an organism have a positive test for acid from glucose in a fermentation tube, but have a negative methyl red test, which is also a glucose fermentation test?

4. Were there any organisms that did not ferment any sugars? If yes, which organisms?

5. When comparing a lactose fermentation tube with a MacConkey plate

   a. what additional information does a fermentation tube give?

   b. what additional information does a MacConkey agar plate give?

EXERCISE

25

Clinical Unknown Identification

Getting Started

In this exercise, you have an opportunity to utilize the knowledge and techniques you have learned in order to identify a mixture of two unknown organisms. You are given a simulated (imitation) **clinical specimen** containing two organisms, and your goal is to separate them into two pure cultures and identify them using different media and tests. The organisms are either associated with disease or are common contaminants found in normal flora or the environment.

Your unknown specimen represents either a urine infection or a wound infection. In actual clinical cases, standardized procedures exist for each kind of specimen. However, you will be identifying only a limited number of organisms. With some careful thought, you can plan logical steps to use in identifying your organisms.

The following are characteristics useful in identifying your unknown organism.

Bacterial Cell Morphology   The size, shape, arrangement, and Gram-staining characteristics of the bacteria as determined by the Gram stain. It also could include the presence of special structures such as endospores.

Colonial Morphology   The appearance of isolated colonies on complete media such as TS agar or blood agar, including their size, shape, and consistency.

Growth on Selective Media   The ability of organisms to grow on **selective media.** Mannitol salt (selects for organisms tolerating 7.5% salt), EMB (selects for Gram-negative organisms), and MacConkey (selects for Gram-negative rods).

*Reactions on **Differential Media***   The color of colonies on eosin methylene blue agar (EMB) or MacConkey agar is based on lactose fermentation (lactose fermenters are purple). The appearance of organisms on mannitol salt agar is based on mannitol fermentation (mannitol fermenters turn medium

yellow). The presence of hemolysis on blood agar constitutes another type of reaction of bacterial enzymes on red blood cells.

Biochemical Capabilities   These capabilities include the ability to ferment different carbohydrates and the production of various end products, as well as the formation of indole from tryptophan and tests such as methyl red, Voges-Proskauer, citrate utilization, urease, catalase, oxidase, and coagulase.

Approach the identification of your "unknown" clinical specimen with the following steps:

1. Make a Gram stain of the specimen.
2. Streak the broth on a complete medium and a selective medium that seem appropriate.
3. After incubation, identify two different colony types and correlate with their Gram reaction and shape. Also correlate the growth and appearance of the colonies on selective media with each of the two organisms.
4. Restreak for isolation. It is useless to do any identification tests until you have pure cultures of the organisms.
5. After incubation, choose a well-isolated colony and inoculate a TS agar slant to be used as your stock culture. Prepare a stock culture for each organism.
6. Inoculate or perform various tests that seem appropriate. Keep careful records. Record your results on the worksheets as you observe them.
7. Identify your organisms from the test results.

Definitions

Clinical specimen.   Cultures encountered in a medical laboratory.

Differential media.   Media that permit the identification of organisms based on the appearance of their colonies.

Selective media. Media that permit only certain organisms to grow and aid in isolating one type of organism in a mixture of organisms.

Objectives

1. To give you an opportunity to apply your knowledge to a microbiological problem.
2. To give you insight into the procedures used to isolate and identify clinical specimens.
3. To teach you to be aware of the presence of contaminants or nonpathogens in clinical specimens.

Note: See exercises 22 and 24 for more information on these organisms and tests.

Materials

First Session

Unknown mixture labeled with hypothetical source (for each student or team of two students)

    Blood agar plate or TS agar plate, 1 per student

    MacConkey agar plate (or EMB agar plate)

    Mannitol salt agar plate

Second and Third Sessions

    TS agar plates

    Nutrient agar slants

    Citrate agar slants

    Urea slant

    Glucose + bromcresol purple agar slants

    Fermentation broths of glucose, lactose, sucrose

    MR-VP broth for the Voges-Proskauer and methyl red test

    Tryptone broth for the indole test

    Kovacs reagent

    Voges-Proskauer reagents A and B

    Methyl red for methyl red test

    Plasma for coagulase test

    Staining material for endospores and capsules

    H_2O_2

References

Difco manual of dehydrated culture media, 10th ed. Detroit: Difco Laboratories, 1984.

Holt, John G. et al. *Bergey's manual of determinative bacteriology.* Baltimore: Williams & Wilkins, 1994.

Nester et al. *Microbiology: A human perspective*, 4th ed., 2004. Chapter 25 and Chapter 27.

Procedure

First Session

1. Make a Gram stain of the broth culture. Observe it carefully to see if you can see both organisms. You can save the slide and observe it again later if you have any doubts about it. You can also save the broth, but one organism may overgrow the other.
2. Inoculate a complete medium agar plate such as TS agar or blood agar, and appropriate selective and differential agar plates. Use MacConkey agar (if you suspect the possibility of a Gram-negative rod in a urine specimen) or a mannitol salt agar plate (if you suspect *Staphylococcus* in a wound infection). Streak the plates for isolated colonies.
3. Incubate at 37°C.

Second Session

1. Examine the streak plates after incubation and identify the two different colony types of your unknown organisms either on the complete medium or the selective media, wherever you have well-isolated colonies. Gram stain each colony type (organisms usually stain better on nonselective media). Also identify each colony type on the selective and differential media so that you know which organisms can grow on the various media. Record their appearance on the differential media as well. It is helpful to circle colonies that you Gram stain on the bottom of the petri plate with a marking pen.
2. Restreak each organism on a complete medium (instead of selective media) for isolation. This technique ensures that all organisms will grow and you will be able to see if you have a mixed culture. Do not

discard your original streak plates of your isolates but store at room temperature. If at some point your isolate does not grow, you will be able to go back to the old plates and repeat the test.

Third Session

1. Observe the plates after incubation. If your organisms seem well isolated, inoculate each one on a TS agar slant to use as your stock culture. If you do not have well-isolated colonies, restreak them. It is essential that you have a pure culture. Possible steps in identifying Gram-positive cocci follow.

2. Look at the possible list of organisms and decide which ones you might have based on the information you have found so far.

These are a few of the test results. Your instructor may provide more. Plan work carefully and do not waste media using tests that are not helpful. For example, a urea slant would not be useful for distinguishing between *Staphylococcus epidermidis* and *S. aureus*. Possible organisms included in unknowns:

Simulated wounds
    Staphylococcus epidermidis
    Staphylococcus aureus
    Micrococcus luteus
    Pseudomonas aeruginosa
Simulated urine infection
    Escherichia coli
    Enterobacter aerogenes
    Proteus
    Enterococcus faecalis
    (plus wound organisms)

This is just a partial list. Others can be added.

Gram-positive Cocci	
Staphylococcus aureus Found in either urine or wounds	Gram-positive cocci in bunches, catalase positive, yellow colony, ferments glucose (acid) and mannitol, coagulase positive, salt tolerant.
Staphylococcus epidermidis Contaminant	Gram-positive cocci in bunches, catalase positive, ferments glucose (acid) but not mannitol, coagulase negative, salt tolerant.
Micrococcus Contaminant	Gram-positive cocci in packets, catalase positive, yellow colony, does not ferment glucose or mannitol, coagulase negative, salt tolerant.
Enterococcus faecalis Found in urine	Gram-positive cocci in chains, catalase negative, ferments glucose (acid), coagulase negative.

Gram-positive Rod	
Bacillus	Large, gram-positive rods, forms spores, catalase positive.

Gram-negative Rods	
Escherichia coli Urine	Glucose and lactose positive (acid and gas), indole positive, methyl red positive, Voges-Proskauer negative, citrate negative, urea negative, oxidase negative.
Proteus Urine	Lactose negative, oxidase negative, urea positive, ferments glucose (acid), indole negative.
Pseudomonas Urine and wounds	Lactose negative, glucose negative, urea negative, oxidase positive, indole negative.
Enterobacter	Glucose and lactose positive (acid and gas), indole negative, methyl red negative, Voges-Proskauer positive, citrate positive, urea negative, oxidase negative.

NOTES:

EXERCISE

25

Laboratory Report: Worksheet and Final Report: Clinical Unknown Identification

Gram stain of original specimen _____

_____

(describe cell shape, arrangement, and Gram reaction)

Gram stains of TS agar subcultures _____

_____

(describe cell shape, arrangement, and Gram reaction)

Test	Organism #1	Organism #2
Colony description		
(Trypticase soy agar or blood)		
Gram stain		
Colony appearance MacConkey (or EMB)		
Colony appearance mannitol salt agar		
Special stains		
capsule		
endospore		
Lactose fermentation		
Glucose fermentation		
Sucrose fermentation		
Mannitol fermentation		

Test	Organism #1	Organism #2
Indole production		
Methyl red		
Voges-Proskauer		
Citrate utilization		
Urea hydrolysis		
Motility		
Catalase test		
Coagulase		
Oxidase		

Final identification:

Final Report

1. What is the identification of your organisms? Discuss the process of identification (reasons for choosing specific tests, any problems, and other comments).

   Organism #1:

   Organism #2:

NOTES:

*I*NTRODUCTION

to Some Immunological Principles and Techniques

Immunology, the study of the body's immune response, is responsible for protecting the body against disease. It is often triggered when foreign substances or organisms invade the body. Examples include pathogenic microbes and chemical compounds they produce, such as foreign materials called **antigens.** In some diseases such as AIDS and cancer, the body's immune response is either seriously weakened or destroyed, whereas for milder diseases the immune response remains complete.

Different forms of the immune response include **phagocytic cells** such as white blood cells (WBCs), **enzymes** such as **lysozyme,** and **antibodies.** WBCs and enzymes are examples of natural immunity since they are already present in the body and need not be triggered by the antigen. In contrast, antibodies are an example of acquired immunity since their formation is triggered only in the presence of the antigen. For some people, the immune response can be triggered by the body's own proteins. This can result in the formation of **autoimmune diseases** such as rheumatoid arthritis and glomerulonephritis.

Phagocytic cells and enzymes are also examples of nonspecific immunity since they can react with a variety of different foreign substances (for instance, phagocytic cells can engulf both inanimate and animate particles). Conversely, antibodies represent a form of specific immunity because they are produced in response to particular antigens (an antibody produced against *Salmonella* cell walls will not react with *Proteus* cell walls).

The exercises include examples of both natural immunity (exercises 26 and 27) and acquired immunity (exercises 28–31). In exercise 26, you study human blood cells and learn how to determine which ones are phagocytic. In exercise 27, you learn how to determine the bacteriostatic activity of the enzyme lysozyme, which occurs naturally in phagocytic white blood cells, saliva, nasal secretions, and tears. Lysozyme is able to digest the cell walls of many bacteria.

The adaptive immunity exercises demonstrate examples of antigen-antibody reactions, which vary from one another depending on the nature of the antigen. If the antigen is particulate (as in cells or insoluble substances such as cardiolipin) an agglutination, or clumping, reaction will occur that can be observed visually (exercise 28). For nonparticulate antigens, a precipitin reaction occurs that can often be observed visually without performing additional visualization procedures (for example, in exercise 31 the use of an agar immunodiffusion for *Coccidioides* identification, in which precipitin lines form in the agar at the site of the antigen-antibody reaction).

Exercises 29 and 30 represent the current state-of-the-art techniques for demonstrating particulate antigen-antibody reactions. One such reaction, latex agglutination, is commonly used for *S. aureus* detection. In exercise 29 you use it for Lancefield grouping of pathogenic streptococci. For this test, serum antibodies are absorbed on the surface of latex beads. When the specific antibody reacts with the *Streptococcus* antigen in question, latex particle aggregation becomes large enough to be seen visually. The other particulate antigen-antibody reaction, ELISA (enzyme-linked immunosorbent assay) is described in exercise 30, where you use it for *Coccidioides* identification. It is also widely employed in other areas such as plant and animal virus identification and for detecting the presence of HIV antibodies.

Definitions

Antibody.   A protein produced by the body in response to a foreign substance (e.g., an antigen), which reacts specifically with that substance.

Antigen.   Any cell particle or chemical that can cause production of specific antibodies and combine with those antibodies.

Autoimmune disease. An immune reaction against our own tissues.

Enzyme. A protein that acts as a catalyst. A catalyst is a substance that speeds up the rate of a chemical reaction without being altered or depleted in the process.

Lysozyme. An enzyme that degrades the peptidoglycan layer of the bacterial cell wall.

Phagocytic cells. Cells that protect the host by ingesting and destroying foreign particles such as microorganisms and viruses.

Some of our cells, although they are part and parcel of us, have not even fixed coherence within our 'rest'. Such cells are called 'free'....The cells of our blood are as free as fish in a stream. Some of them resemble in structure and ways so closely the little free swimming amoeba of the pond as to be called amoeboid. The pond amoeba crawls about, catches and digests particles picked up in the pond. So the amoeboid cells inhabiting my blood and lymph crawl about over and through the membranes limiting the fluid channels in the body. They catch and digest particles. Should I get a wound they contribute to its healing. They give it a chance to mend, by eating and digesting bacteria which poison it and by feeding on the dead cells which the wound injury has killed. They are themselves unit lives and yet in respect to my life as a whole, they are components in that corporate life.

Sherrington, Man or His Nature

EXERCISE

Differential White Blood Cell Stains

Getting Started

This exercise deals with the cellular forms of the immune system, specifically the white blood cells. For the most part, they can be distinguished from one another using a blood smear stained with a differential stain such as Wright's stain. This stain uses a combination of an acid stain such as eosin and a basic stain such as methylene blue. They are contained in an alcoholic solvent (methyl alcohol) which fixes the stains to the cell constituents, particularly since the basophilic granules are known to be water soluble. With this stain, a blood smear shows a range in color from the bright red of acid material to the deep blue of basic cell material. In between are neutral materials that exhibit a lilac color. There are also other color combinations depending upon the pH of the various cell constituents.

The two main groups of WBCs are the granulocytes (cytoplasm which contains granules) and the agranulocytes (clear cytoplasm). The granulocytes are highly phagocytic and contain a complex, segmented nucleus. The agranulocytes are relatively inactive and have a simple nucleus or kidney-shaped nucleus. Common cell types found in the granulocytes are **neutrophils, eosinophils,** and **basophils.** The basic agranulocyte cell types are the **lymphocytes** and **monocytes.** Another white cell type found in blood is the **platelet** (very small, multinucleate, irregular pinched off parts of a **megakaryocyte**). Platelets aid in the prevention of bleeding. The appearance of these cell types in blood stained with a differential stain are illustrated in figure 26.1.

Differential blood stains are important in disease diagnosis, since certain WBCs either increase or decrease in number, depending on the disease. In making such judgments, it is important to know the appearance of normal blood (color plate 25). The microscopic field shown includes mostly RBCs with a few neutrophils, only one lymphocyte, and some platelets.

Figure 26.1 Blood cell types present in human peripheral blood. The granular leukocyte names find their origin from the color reaction produced by the granules after staining with acidic and basic components of the staining solution. Neutrophil = neutral-colored granules; basophil = basic color; and eosinophil = acid color.

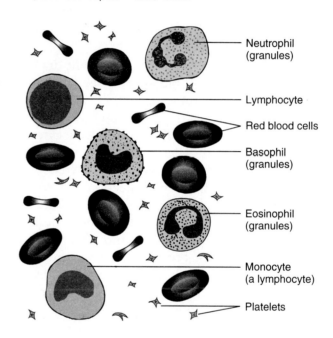

Neutrophil (granules)

Lymphocyte

Red blood cells

Basophil (granules)

Eosinophil (granules)

Monocyte (a lymphocyte)

Platelets

A quantitative description of the various cell types found in normal blood is shown in table 26.1. The red blood cells (RBCs), which are also called erythrocytes, make up the largest cell population. RBCs constitute an offensive weapon because they transport oxygen to various body parts as well as

Table 26.1 Cellular Description of Normal Blood*

Total Counts	Differential WBC Counts
RBC 5,200,000/ml	Neutrophils 64%
WBC 7,200/ml	Leukocytes 33%
Platelets 350,000/ml	Monocytes 2%
Basophils 50/ml	Eosinophils 1%

*From Kracke (see References)

break down carbon dioxide to a less toxic form. The red blood cells in humans and all other mammals (except members of the family Camelidae, such as the camel) are biconcave, circular discs without nuclei.

RBCs are produced in the red bone marrow of certain bones. As they develop, they produce massive quantities of hemoglobin, the oxygen transporting pigment that contains iron, which in the oxygenated form gives blood its red color. Worn-out RBCs are broken down at the rate of 2 million cells per second in the liver and spleen by phagocytic white blood cells (WBCs). Some of the components of the RBCs are then recycled in order for the body to maintain a constant number of RBCs in the blood.

The WBCs, or leukocytes, represent approximately $\frac{1}{800}$ of the total blood cells (see table 26.1). They are defensive cells, specialized in defending the body against infection by microorganisms and other foreign invaders. Many of the leukocytes are **amoeboid,** capable of moving independently through the bloodstream. They also move out into the tissues where they repel infection and remove damaged body cells from bruised or torn tissue.

In this exercise, you have an opportunity to prepare, stain differentially, and observe some human blood slides. For this purpose, the blood used can be outdated whole blood obtained from a blood bank. If this is not available, commercially prepared blood slides are available.

Definitions

Amoeboid. To make movements or changes in shape by means of protoplasmic flow.

Basophil. A granulocyte in which the cytoplasmic granules stain dark purplish blue with methylene blue, a blue basophilic-type dye found in Wright's stain.

Eosinophil. A granulocyte in which the cytoplasmic granules stain red with eosin, a red acidophilic type dye found in Wright's stain.

Lymphocyte. A colorless agranulocyte produced in lymphoid tissue. It has a single nucleus with very little cytoplasm.

Megakaryocyte. A large cell with a lobulated nucleus that is found in bone marrow, and is the cell from which platelets originate.

Monocyte. A large agranulocyte normally found in the lymph nodes, spleen, bone marrow, and loose connective tissue. It is phagocytic with sluggish movements. When stained with Wright's stain, it is difficult to differentiate from a junior neutrophil.

Neutrophil. A mature granulocyte present in peripheral circulation. The cytoplasmic granules stain poorly or not at all with Wright's stain. The nuclei of most neutrophils are large, contain several lobes, and are described as polymorphonuclear (PMN) leukocytes.

Plasma. The fluid portion of the blood, exclusive of cells, before clotting.

Platelet. A small oval to round colorless biconcave disc, 3 microns in diameter. Plays a roll in clotting of blood.

Objectives

1. To provide historical and background information on blood and some of its microscopic cell types, their origin, morphology, number, and role in fighting disease.
2. To prepare two stained blood slides: the first for use in observing the cellular appearance of normal blood, and the second slide for use in determining a differential WBC count.

References

Johnston, R. "Monocytes and macrophages." *New England Journal of Medicine* 318 (no. 12): 747–752, 1988.

Kracke, Roy R. *Diseases of the blood and atlas of hematology,* 2nd ed. Philadelphia: J. B. Lippincott Co., 1941. Excellent sourcebook, beautifully illustrated, containing clinical and hematologic descriptions of the blood diseases, and a section on technique and terminology. Unfortunately, this is the latest edition.

Lechevalier, H. A., and Solotorovsky, M. *Three centuries of microbiology.* New York: McGraw-Hill Book Co., 1965. Contains historical information on the immune response.

Metschnikoff, Élie. 1884. "A disease of *Daphnia* caused by a yeast." In *Milestones in microbiology,* translated by Thomas Brock, Washington, D.C.: American Society for Microbiology,

1961. Contains historical information on the immune response.

Nester et al. *Microbiology: A human perspective,* **4th ed., 2004. Chapter 15, Introduction.**

Materials

Either outdated blood bank whole blood or prepared commercial unstained or stained human peripheral blood smears. For student use, outdated whole blood should be dispensed with either a plastic dropper or a dropping bottle capable of dispensing a small drop.

Note: In the event of spilled blood, use disposable gloves and towels to remove blood. Then disinfect the area with a germicide such as hypochlorite bleach diluted approximately 1:20 with water. See Nester et al., Section 5.4 for additional comments.

For use with whole blood:

New microscope slides, 3

Plastic droppers for dispensing blood on slides, dispensing Wright's stain, and adding phosphate buffer

Hazardous waste container for droppers and slides

For use with whole blood and unstained prepared slides:

Wright's stain, dropping bottle, 1 per 2 students

Phosphate buffer, pH 6.8, dropping bottle, 1 per 2 students

Wash bottle containing distilled water, 1 per 2 students

A Coplin jar with 95% ethanol

Staining rack

Colored pencils for drawings: pink, blue, purple, or lavender

---

Procedure

Safety Precautions: When using whole blood, be careful not to dispense on surfaces other than slides. Wipe up any spilled blood immediately and disinfect area since blood residues promote the growth of microbial contaminants. Wash and rinse hands immediately after preparing slides for staining.

1. Prepare three clean microscope slides free of oil and dust particles as follows:
   a. Wash slides with a detergent solution, rinse thoroughly.
   b. Immerse slides in a jar of 95% alcohol.
   c. Air dry and polish with lens paper.
2. Place a drop of blood on one end of a clean slide (figure 26.2a). Repeat with a second clean slide.
3. Spread the drop of blood on the slide as follows:
   a. Place the slide on your laboratory bench top. With your thumb and middle finger, firmly hold the sides of the slide on the end where the drop of blood is located.
   b. With your other hand, place the narrow edge of a clean slide approximately ½" in *front* of the drop at an angle of about 30° (figure 26.2b).
   c. Carefully push the spreader slide *back* until it comes in contact with the drop, at which point the drop will spread outward to both edges of the slide (figure 26.2c).
   d. Immediately with a firm steady movement push the blood slowly toward the opposite end of the slide (figure 26.2d).

   Note: Use of the above procedural restraints (a small drop, a small spreader slide angle, and a *slow*, steady spreader slide movement) should provide a thin film for study of red cells. A good smear has the following characteristics: smooth, without serrations; even edges; and distributed uniformly over the middle ⅔ of the slide.
   e. Allow slide to air dry for 5 minutes. Do not blot.
   f. For the second slide, prepare a thicker film by using a larger spreader slide angle (45°), and by pushing the blood more rapidly to the opposite end of the slide. The second slide is best for determining the differential white blood cell count.

   Note: The unused end of the first spreader slide can be used to prepare the second slide. Discard used spreader slide in the hazardous waste container.

Figure 26.2 (a–d) Method for preparing a blood smear.

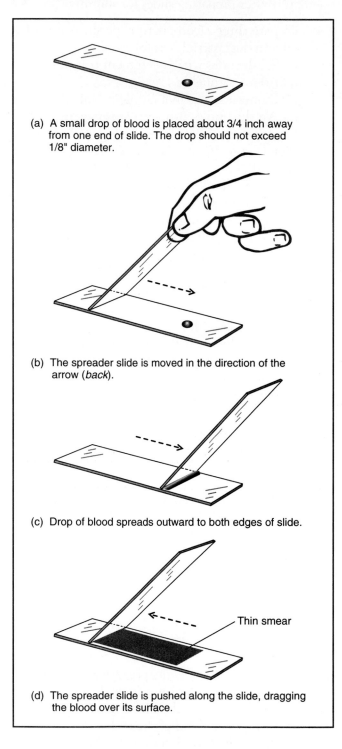

(a) A small drop of blood is placed about 3/4 inch away from one end of slide. The drop should not exceed 1/8" diameter.

(b) The spreader slide is moved in the direction of the arrow (*back*).

(c) Drop of blood spreads outward to both edges of slide.

Thin smear

(d) The spreader slide is pushed along the slide, dragging the blood over its surface.

4. Stain the blood smears with Wright's stain as follows:
   a. Suspend the slides such that they lie *flat* on the staining rack supports.
   b. Flood or add 15 drops of Wright's stain to each blood smear. Let it stand for 3 to 4 minutes. This fixes the blood film.
   c. Without removing stain, add an equal volume of phosphate buffer. Blow gently through a pipet on each side of the slide to help ensure mixing of stain and buffer solutions.
   d. Let stand until a green, metallic scum forms on the surface of the slide (usually within 2 to 4 minutes).
   e. Wash off the stain with water. Begin washing while the stain is on the slide in order to prevent precipitation of scumlike precipitate which cannot be removed. The initial purple appearance on the slide should be washed until it is a lavender-pink.
   f. Wipe off excess stain from the back of the slide and allow it to stand on end to dry (which is preferable to drying between bibulous paper).
5. Examine stained blood smears:
   a. Make an initial examination of the first blood smear with the low power objective to find the most suitable areas for viewing with the oil immersion objective.
   b. Next using the oil immersion lens, make a study of the various WBC types present: basophils, eosinophils, lymphocytes, monocytes, neutrophils, and platelets. For help in this study, consult color plate 23, the Definitions section describing their staining characteristics with Wright's stain, and figure 26.1. Prepare color drawings of your findings on the Laboratory Report sheet.
   c. Conduct a differential white blood cell count using the second blood smear. For normal blood with a leukocyte count of 5,000 to 10,000 WBCs/ml, one would classify 100 leukocytes. In order to do this, you may have to examine the number and kinds of WBCs present in as many as twenty microscopic fields. Record your findings in table 26.2 of the Laboratory Report and calculate the percentage of each WBC type.

EXERCISE

Laboratory Report: Differential White Blood Cell Stains

Results

1. Color drawings of RBCs and various WBCs found in blood smears stained with Wright's stain.

   RBCs   Neutrophils   Eosinophils

   Basophils   Monocytes   Lymphocytes

2. Differential WBC count. In table 26.2, record the kinds of leukocytes found as you examine each microscopic field. After counting 100 WBCs, calculate their percentages from the totals found for each type. Also record the number of microscopic fields examined to find 100 WBCs: _____

Table 26.2   Kinds and Percentages of WBCs Found in Blood Smear

Neutrophils	Eosinophils	Basophils	Lymphocytes	Monocytes
Total				
Percent				

Questions

1. What problems if any did you find in preparing and staining your blood smears? Indicate any differences noted between thin and thick smears.

2. Were your blood stains satisfactory? Did the stained cells resemble those in figure 26.1 and color plate 23? Were they better?

3. Did your differential white blood cell count percentages compare with the percentages in normal blood (table 26.1)? If not, give an explanation.

4. Were there any WBC types that you did not find in your blood smear? If so, which one(s)? Why did you not find them?

5. Matching (you may wish to consult your text).

    a. Neutrophils   ____ Involved in antibody production
    b. Basophils   ____ A minor phagocytic cell
    c. Monocytes   ____ Increased number in parasitic infections
    d. Eosinophils   ____ Largest WBC
    e. Lymphocytes   ____ Inflammatory WBC

EXERCISE

Lysozyme, an Enzymatic Form of Natural Resistance

Getting Started

A number of antimicrobial chemicals have been isolated from animal cells and body fluids. Among these are two important proteins: **lysozyme** and **complement,** both of which are examples of **natural immunity.** Complement is necessary for certain antigen-antibody reactions in which it becomes fixed to the antigen-antibody complex. For example, the Wassermann complement fixation test is used for syphilis diagnosis. It was developed by Bordet and Gengou in 1901. Through textbook reading, you should become familiar with the principles of this classic syphilis test. (See **Nester et al. *Microbiology: A human perspective,* 4th ed., 2004. Chapter 17, Section 17.8.**)

Lysozyme is a **proteolytic enzyme** found in saliva and nasal secretions. It is also found in phagocytic WBCs (see exercise 26), where it functions as a biocide, and in egg white. Enzymatically, it is able to degrade the peptidoglycan layer of the bacterial cell wall, thereby weakening and eventually destroying it. It is particularly active against Gram-positive bacteria that have an exposed peptidoglycan layer. Experimentally, when bacterial cells are treated with lysozyme in a solution of higher osmotic pressure, such as 0.6 M sucrose, the cell wall dissolves but cell lysis is inhibited (see exercise 13).

Chemically, lysozyme is a **globulin protein** (N-acetyl-muramide hydrolase) that was discovered by Alexander Fleming in 1922. It functions by dissolving the **peptidoglycan layer** of the bacterial cell wall (see **Nester et al. *Microbiology: A human perspective,* 4th ed., 2004. Chapter 3, Section 3.6** for more details). It also acts on **chitin,** which is the principal component in the exoskeleton of molds, yeasts, invertebrates, and arthropods. Another source of lysozyme is the tail section of the bacteriophage. Lysozyme helps initially with phage entry through the bacterial cell wall. Later phage stages direct the host cell to produce more lysozyme for use in facilitating release of new phage particles.

In this exercise, you have an opportunity to assay the antimicrobial activity of lysozyme collected from tears and egg white, and to compare their antimicrobial activity with a commercial lysozyme preparation of known activity.

Definitions

Chitin. A polysaccharide structurally similar to cellulose.

Complement. A system of at least 26 serum proteins that act in sequence, producing certain biological effects concerned with inflammation and the immune response.

Globulin protein. A class of simple proteins characterized by their almost complete insolubility in water, solubility in dilute salt solutions, and coagulability by heat. They occur widely in plant and animal tissues as blood plasma or serum.

Lysozyme. An enzyme able to attack and destroy bacterial cell walls. It occurs naturally in tears, saliva, phagocytic WBCs, and egg white.

Natural immunity. An immunity to infectious disease in a species occurring as a part of its natural biologic makeup.

Peptidoglycan layer. The rigid backbone of the bacterial cell wall, composed of repeating subunits of N-acetylmuramic acid and N-acetylglucosamine and other amino acids.

Proteolytic enzyme. An enzyme able to hydrolyze proteins or peptides with the resulting formation of simpler, more soluble products such as amino acids.

Objectives

1. To become more familiar with the process of natural immunity, and how chemicals such as complement and lysozyme function in immunity.
2. To learn how to determine the antimicrobial activity of various natural lysozyme preparations.

References

Fleming, A. 1922. Proc. of the Royal Soc. of London, Ser. B, 93, 306.

Nester et al. *Microbiology: A human perspective,* 4th ed., 2004. Chapter 3, Section 3.6.

Osserman, et al., eds. *Lysozyme.* New York: Academic Press, 1974.

Materials

Per pair of students

Nutrient broth cultures (37°C, 24 hour) of *Staphylococcus epidermis* and *E. coli*

Sabouraud's dextrose broth culture (25°C, 24 hour) of *Saccharomyces cerevisiae*

Melted nutrient agar deeps, 2, and Sabouraud's dextrose agar deep, 1, held in a 48°C water bath

Sterile petri dishes, 4

Petri dish containing 9 sterile filter paper discs (approximately ½" diameter)

Petri dish containing 1–2 ml of aseptically prepared raw egg white

Petri dish containing 1–2 ml of lysozyme chloride (Sigma cat. # L-2879) with activity of approximately 60,000 units per mg of protein. Diluted 1:10 with sterile distilled water

Test tube containing 9 ml sterile distilled water

Raw onion

Scalpel or sharp knife for cutting the onion

Mortar and pestle

Pair of tweezers

Sterile 1-ml pipets, 4

Ruler calibrated in mm

Procedure

First Session

1. With a glass-marking pencil, divide the underside of 3 petri dishes in thirds. Label one part T for tears, another part EW for egg white, and the remaining part L for lysozyme.

2. Label each petri dish with the name of one of the three test organisms. Also include your name and the date on the bottom.

3. Prepare pour plates of *Staphylococcus* and *E. coli* by suspending the broth culture. With a sterile 1-ml pipet aseptically transfer a 1-ml aliquot to the respectively marked petri dish. Discard pipet in the hazardous waste container. Add the contents of a tube of melted nutrient agar to the dish, replace cover, and rotate the dish 6 to 8 times on the surface of your desk in order to distribute the cells uniformly. Allow agar to harden. Repeat above procedure for the other culture.

4. Repeat step 3 above for *S. cerevisiae* using a melted Sabouraud's agar deep.

5. In order to induce tear secretions, one student should remove the outer skin from an onion. Then cut the onion into small pieces and crush in a mortar with a pestle. The other student should be prepared to collect the secreted tears in a sterile petri dish—0.5 to 1 ml is sufficient.

6. Alcohol sterilize tweezers and air cool. Aseptically remove a filter paper disc and dip it into the tears. Transfer the moistened disc to the center of the area marked T on the *S. epidermis* agar plate.

7. Repeat step 6 by transferring moistened tear discs to petri dishes containing the remaining two test organisms.

8. Repeat steps 6 and 7 with the egg white preparation.

9. Repeat steps 6 and 7 with the lysozyme preparation.

10. Incubate the petri dishes containing *Staphylococcus* and *E. coli* at 37°C for 24 hours and the petri dish containing *S. cerevisiae* at 25°C for 48 hours.

Second Session

1. Observe the petri dishes for zones of inhibition around the filter paper discs. With a ruler, determine their diameter in mm and record your results in table 27.1 of the Laboratory Report.

Name _____ Date _____ Section _____

EXERCISE

Laboratory Report:
Lysozyme, an Enzymatic Form of Natural Resistance

Results

1. Complete table 27.1 (see instructions in Second Session, Procedure step 1).

Table 27.1   Antimicrobial Activity of Various Lysozyme Extracts

Test Organism	Diameter of Inhibition Zone (mm)		
	Tears	Egg White	Lysozyme*
S. epidermis			
E. coli			
S. cerevisiae			

*1:10 dilution

2. Which of the three preparations was the most active?_____Least active?_____Consider the lysozyme *dilution* when preparing your answer.

3. Which organism(s) were not inhibited by lysozyme?

4. Knowing that the above lysozyme preparation has an activity of 60,000 units per mg of protein, calculate the lysozyme activity in similar units for tears and egg white against *S. epidermis* and *E. coli*. Some of your results from table 27.1 are necessary for making this calculation. Record your findings in table 27.2.

Table 27.2   Units of Lysozyme Activity for Egg White and Tears

Test Organism	Units of Lysozyme Activity/Mg of Protein	
	Egg White	Tears
S. epidermis		
E. coli		

Questions

1. What similarity do lysozyme and penicillin have in their mode of antimicrobial action? How do they differ?

2. Why must tweezers be cool before dipping the filter paper discs in the lysozyme solutions?

3. What large groups of microorganisms are susceptible to lysozyme? Resistant to lysozyme?

4. Procedurally, what additional important step would be necessary to evaluate the lysozyme activity of nasal secretions? Why?

5. Why are most Gram-negative bacteria not lysed by lysozyme, yet they have a peptidoglycan cell wall structure similar to that of the Gram-positive bacteria?

NOTES:

EXERCISE

Traditional Agglutination Reactions Employing Microbial and Nonmicrobial Antigens

Getting Started

An **agglutination** reaction is an **antigen-antibody** reaction in which the antigen consists of particulate matter such as cells or synthetic material and the antibodies are described as **agglutinins.** The usefulness of this simple procedure—mixing antigen particles with antibody—gave rise to the era of **serodiagnosis,** advanced understanding of the role microorganisms play in causing disease. It eventually led to the discovery of the ABO blood groups.

Agglutination occurs in two steps: the specific combination of antigen and antibody, which is then followed by the visible aggregation of the particles. Factors such as the charge of the antigen-antibody particles, buffering, and viscosity of the test medium play a role in proper agglutination of antigen particles by antibodies. A disadvantage of the agglutination phenomenon is that the reaction is semiquantitative (only accurate to a fourfold difference in antibody **titer**). However, the facts that numerous systems lend themselves to agglutination reactions, the basic simplicity of agglutination systems developed to date, and the high sensitivity of agglutination-based reactions encourage wide use of such tests.

Use of Microbial Antigens for Diagnosis of Infectious Diseases

Some of the most useful agglutination tests are used for infectious disease diagnosis, such as for salmonellosis and rickettsiosis. Early identification of initially high or rising agglutinin titers to these organisms offers a powerful laboratory adjunct to clinical diagnosis.

For example, the Widal test, which was devised in 1954, is used to identify serum antibodies against various *Salmonella* species antigens, which vary in composition from body parts (**somatic** and capsular antigens) to flagellar antigens. A titer of somatic and flagellar antibodies equal to a serum dilution of 1/80 or greater suggests an active infection (see figure 28.2).

Another useful agglutination test is the Weil-Felix test, which will be evaluated in this exercise. It uses *Proteus* antigens to detect **cross-reacting** rickettsial serum antibodies. Rickettsial diseases diagnosed most frequently in the United States include murine typhus fever, Q fever, and Rocky Mountain spotted fever. The latter is caused by *Rickettsia rickettsii*, which is transmitted to humans by a tick **vector.** Thus, the Weil-Felix reaction is useful in screening clinically suspect patients for high titers of *Proteus* agglutinins for presumptive diagnosis of Rocky Mountain spotted fever and murine typhus fever (whose vector is infected flea feces). *Proteus* antigens are used because the two rickettsia share some minor antigens with these bacteria.

Table 28.1 shows that *Proteus vulgaris* strains OX-19 and OX-2 are agglutinated by sera of individuals infected with most members of the spotted fever group. Rickettsial antibodies can reach detectable levels within one week after the onset of symptoms, and usually reach a maximum titer within a few months.

Table 28.1	Weil-Felix Reactions in Rickettsioses*		
Disease	OX-19	OX-2	OX-K
Epidemic typhus	++++	+	0
Murine typhus	++++	+	0
Brill-Zinsser disease	variable	variable	0
Spotted fever[a]	++++	+	0
	+	+++	0
Rickettsial pox	0	0	0
Scrub typhus	0	0	+++
Q fever	0	0	0
Trench fever	0	0	0

[a]Spotted fever immune sera can agglutinate either OX-19 or OX-2, or both.
*From the *Manual of Clinical Immunology*, p. 709, 1980.
National Committee for Clinical Laboratory Standards. *Performance Standards for Antimicrobial Disk Susceptibility Tests.* Fifth Edition; Approved Standard. Copyright © 1993 The National Committee for Clinical Laboratory Standards. By permission.

Although the Weil-Felix test is the least sensitive of immunological tests currently available for use in rickettsial disease diagnosis, it is used in this exercise to illustrate a classical agglutination reaction. It is also the least specific test since the *Proteus* organisms that are used are also agglutinated by antibodies occurring as a result of *Proteus* urinary tract infections.

Antigens available for the Weil-Felix test are suspensions of whole, unflagellated *P. vulgaris* strains OX-19 and OX-2, and *P. mirabilis* strain OX-K. Motility of the organisms should be checked with a hanging drop culture since flagellar H antigens, if present, prevent agglutination of the somatic O antigens by rickettsial antibodies. Controls for the W-F test should include a positive human antirickettsial serum and a normal human serum (negative control).

The W-F test can be performed by both rapid slide and tube agglutination methods. The former is a qualitative method and the latter a quantitative method. When conducting the test, keep in mind that it is not specific for rickettsial infections. Agglutination may occur from a previous *Proteus* infection and also from other cross-reacting organisms. However, if properly performed, the test remains a useful tool in the diagnosis of rickettsial disease, due to wide availability of antigens.

Use of a Nonmicrobial Antigen for Initial Syphilis Diagnosis

Tests for syphilis diagnosis involve two generations of antigens: the first was **cardiolipin** (derived from beef heart) and the second was the infectious disease organism *Treponema pallidum*. It is paradoxical that the use of cardiolipin still remains superior to the use of *T. pallidum* even though it is biologically nonspecific. In this test, cardiolipin antigens are used to detect antilipid antibodies, traditionally termed **reagin,** in the patient's serum. It is not known whether these antibodies are invoked by lipid antigen present in *T. pallidum* or by the host-parasite reaction.

The nontreponemal tests most frequently used in the United States today are the Venereal Disease Research Laboratory (VDRL) test and the rapid plasma reagin (RPR) 18 mm circle card test. Both are simple, rapid, reproducible, and inexpensive; both can be used either qualitatively or quantitatively.

The RPR card test has various advantages over the VDRL test: it is available as a kit test containing all needed reagents and controls including a prepared antigen suspension; unheated serum is used; the reaction is read macroscopically; and most materials are throwaways.

The unheated serum reagin (USR) test to be evaluated in this exercise is a cross between the VDRL and RPR syphilis diagnosis tests. The USR test, like the RPR test, uses unheated serum. However, unlike the RPR test and like the VDRL test, agglutination must be observed microscopically rather than macroscopically. All three tests are alike in that they use the nonmicrobial antigen cardiolipin.

The nontreponemal tests are of greatest value when used as screening procedures and for evaluating patient response to syphilis therapy. Treponemal tests are used primarily to determine whether a reactive nontreponemal test is due to syphilis or some other condition. The treponemal tests can also be used to detect syphilis in patients with negative nontreponemal tests but with clinical evidence of syphilis. False-positive treponemal tests should be further evaluated. Unlike the nontreponemal test, the treponemal tests do not indicate the patient's response to treatment, and quantitative tests are of no value in diagnosis or prognosis.

Although the immunological response to infection with *T. pallidum* is complex and poorly understood, serological tests are frequently the only basis for syphilis diagnosis and for evaluating patient response to treatment (Coffey and Bradford, 1980). Nontreponemal tests usually become reactive 4–6 weeks after infection or 1–2 weeks after the first visible chancre appears. The specificity of nontreponemal tests is lacking since they can be reactive in a variety of other conditions. Treponemal tests, although specific, vary in their ability to react in early syphilis. Once reactive, all treponemal tests tend to remain so for years. None of the treponemal tests distinguish between syphilis and other treponemal infections, such as bejel, pinta, and yaws.

Definitions

Agglutination.   The aggregation of foreign cells by antibodies (agglutinins) or by synthetic particles (agglutinogens) into visible clumps.

Agglutinin. A specific antibody capable of affecting the agglutination of the agglutinogen that stimulated its production.

Antibody. A protein produced by the body in response to a foreign substance.

Antigen. A foreign substance that incites production of specific antibodies.

Cardiolipin. A phosphatide obtained from beef heart which is used as an antigen in diagnostic tests for syphilis.

Cross-reacting antibodies. The immunological phenomenon wherein one antigen reacts with antibodies that were developed against another antigen.

Reagin. A substance in the blood of people with syphilis that sometimes functions as an antibody against a syphilis antigen.

Serodiagnosis. Diagnosis of disease by the use of serum as in the USR test for syphilis and the Weil-Felix test for rickettsial diseases.

Somatic antigen. An antigen from the body of a bacterial cell.

Titer. A measure of the serum antibody level. Usually measured as the highest dilution of serum that will test positive for that antibody. The titer is often expressed as the reciprocal of that dilution.

Vector. An agent, often an insect, that transmits an infectious disease from one host to another host. For example, fleas transmit bubonic plague from rats to humans.

Objectives

1. To provide information about agglutination tests using both microbial and nonmicrobial antigens.
2. To describe advantages and disadvantages of these tests when used for diagnosis and for evaluating patient response to therapy. Diseases discussed are Rocky Mountain spotted fever (a rickettsial disease) and syphilis (a treponemal disease).
3. Laboratory evaluation of a conventional agglutination method, the Weil-Felix test, which uses a microbial *Proteus* antigen for

presumptive diagnosis of rickettsial diseases. The test is evaluated qualitatively by the rapid slide method and quantitatively by the tube dilution method.

4. To evaluate an agglutination reaction with the nonmicrobial antigen cardiolipin, which is used for presumptive syphilis diagnosis, the USR (unheated serum reagin) test.

References

Coffey, E., and Bradford, L. "Serodiagnosis of syphilis," 530–541, in the *Manual of clinical immunology* (see below).

D'Angelo, L. J., Winkler, W. G., and Bregman, D. J. 1978. "Rocky Mountain spotted fever in the United States, 1975–77." *Journal of Infectious Diseases* 138:273–276.

Manual of clinical immunology, 2nd ed. Eds. Noel R. Rose and Herman Friedman, Washington, D.C.: American Society for Microbiology, 1980. Contains a wealth of information presented by various authors clearly and concisely. A must for any instructor or student desiring to obtain a solid introduction to clinical immunology. See also 4th ed., 1992.

Manual of tests for syphilis. Venereal Disease Program, 1969. Atlanta: Centers for Disease Control.

Nester et al. *Microbiology: A human perspective*, 4th ed., 2004. Chapter 17, Section 17.4.

Materials

Use of *Proteus* Antigens to Detect Rickettsial Antibodies (Weil-Felix test)

For use with rapid slide and tube test:

Proteus OX-2 or OX-19 antigen and antiserum, for source ask your instructor

Suitable light source for observing agglutination, such as a gooseneck, fluorescent, or fiber optic lamp

A rubber bulb for pipetting serum and antiserum

Isotonic saline (0.85 g NaCl/100 ml distilled water)

For use with rapid slide test:

Clean microscope slides, 3

Applicator sticks or toothpicks

Dropper delivering approximately 0.03 ml

Serological pipets, 0.2 ml, cotton plugged, 2

For use with tube test:

Serology test tubes, approximately 10 × 100 mm, 10

Test tube rack for holding serology test tubes

Sterile 5-ml serological pipets, cotton plugged, 2

Sterile 1-ml serological pipets, cotton plugged, 2

Water bath, 37°C

Use of a Nonmicrobial Antigen (Cardiolipin) for Syphilis Diagnosis

USR antigen and USR test control serum set, for source ask your instructor

Negative control serum

Hypodermic needle without bevel, 18 gauge adjusted to drop ¼₅ ml per drop

Syringe, Luer-type, 1–2 ml

Absolute alcohol and acetone for rinsing syringe with needle

Clean microscope slides, 2

Serological pipets, 0.2 ml, cotton plugged, 3

Ruler calibrated in mm

Procedure

The Rapid Slide Agglutination Test for Detection of Rickettsial Antibodies: A Qualitative Screening Test Employing a Microbial (*Proteus*) Antigen

1. With a glass-marking pencil, mark three *clean* microscope slides with two 16-mm (⅝″) circles per slide.
2. Using a 0.2-ml pipet with a rubber bulb, pipette the following amounts of *Proteus* antiserum into the first five circles: 0.08, 0.04, 0.02, 0.01, and 0.005 ml. Discard used pipet in

hazardous waste container. To the sixth circle add 0.08 ml of 0.85% saline using a fresh pipet (negative control).
3. To each circle, add one drop of *Proteus* antigen with a dropper.

Note: Shake the antigen well before using.

4. Mix each antiserum-antigen composite with an applicator stick or toothpick. Start with the 0.005-ml serum dilution and work back to the 0.08-ml dilution. Discard mixing tool in the hazardous waste container. The final dilutions correspond approximately with the macroscopic tube test dilutions, which are 1:20, 1:40, 1:80, 1:160, and 1:320, respectively.
5. Hold the slide in both hands and gently rotate 15–20 times (figure 28.1).
6. Observe for macroscopic agglutination (clumping) using a suitable light source.

Note: Make observations within 1 minute, since later reactions may be due to drying of reactants on slide.

7. Record the amount of clumping for the various dilutions in table 28.2 of the Laboratory Report as follows:
   - − no agglutination
   - + trace agglutination
   - 1 + approximately 25% cell clumping
   - 2 + approximately 50% cell clumping
   - 3 + approximately 75% cell clumping
   - 4 + complete agglutination

Figure 28.1 Rotation method used to initiate agglutination of antigen-antibody mixture.

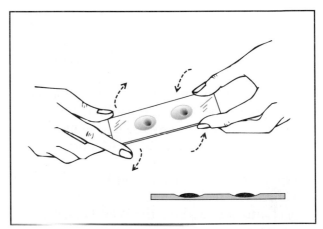

The Tube Dilution Agglutination Test for Detection of Rickettsial Antibodies: A Quantitative Test Employing a Microbial (*Proteus*) Antigen

Note: May be used as a demonstration exercise.

1. Prepare serial dilutions containing 0.5 ml of positive control antiserum in serology test tubes as follows (figure 28.2):
   a. Place ten serology test tubes in a test tube rack.
   b. With a 1-ml serological pipet, transfer 0.9 ml of 0.85% saline into the first tube and 0.5 ml into each of the remaining tubes. Discard pipet.
   c. With a fresh 1-ml serological pipet containing a rubber bulb, add 0.1 ml of antiserum to the first tube. Mix well by drawing in and out of pipet about 7 times. Discard used pipet in hazardous waste container. Use a fresh pipet for preparing each of the remaining dilutions.
   d. Transfer 0.5 ml from tube 1 to tube 2. Mix and transfer 0.5 ml of tube 2 to tube 3, mix; continue above dilution process through tube 9, discarding 0.5 ml from tube 9 into hazardous waste container.

Note: Tube 10, which does not contain serum, serves as an antigen control. The serum dilution in the first tube is 1:10 because 0.1 ml serum was added to 0.9 ml saline, and the serum dilution in tube 2 is 1:20 because 0.5 ml of the 1:10 dilution was added to an equal volume of saline. The same principle holds for tube 3, which is 1:40 or double that of tube 2. This illustrates the principle of how to prepare doubling dilutions.

2. With a 1-ml serological pipet containing a rubber bulb, add 0.5 ml of well suspended *Proteus* antigen to each of the ten tubes. Discard pipet in hazardous waste container.
3. Gently shake the rack to mix antigen and antiserum. The resultant dilutions are 1:20 through 1:5120, respectively.
4. Incubate the rack in a 37°C water bath for 2 hours, followed by overnight incubation in a 2–8°C refrigerator.

Note: It is important to use the recommended time and temperature of incubation, and to make certain that the water bath is in a location free of mechanical vibration.

As you have seen, the exercise was conducted using a positive control antiserum. In a hospital environment, serum drawn from a patient would constitute the unknown test serum. For greater proficiency in test interpretation, it is also important to include a febrile negative control antiserum. The latter is processed in the same manner as the positive control antiserum.

Figure 28.2 Serial tube dilution test protocol employing doubling dilutions.

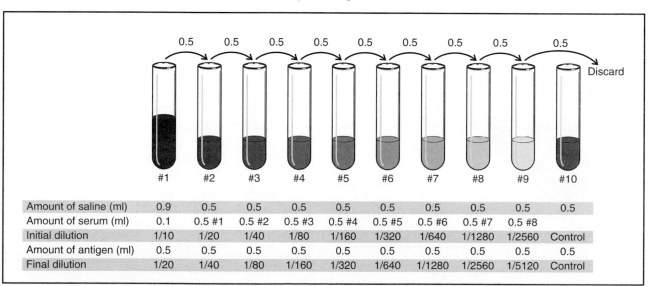

	#1	#2	#3	#4	#5	#6	#7	#8	#9	#10
Amount of saline (ml)	0.9	0.5	0.5	0.5	0.5	0.5	0.5	0.5	0.5	0.5
Amount of serum (ml)	0.1	0.5 #1	0.5 #2	0.5 #3	0.5 #4	0.5 #5	0.5 #6	0.5 #7	0.5 #8	
Initial dilution	1/10	1/20	1/40	1/80	1/160	1/320	1/640	1/1280	1/2560	Control
Amount of antigen (ml)	0.5	0.5	0.5	0.5	0.5	0.5	0.5	0.5	0.5	0.5
Final dilution	1/20	1/40	1/80	1/160	1/320	1/640	1/1280	1/2560	1/5120	Control

5. Examine the tubes *qualitatively* for the type of sediment using a mirror, preferably concave, under a light (figure 28.3).

The negative control tube should show that the antigen has settled out in the bottom of the tube in a small, round disc with smooth edges (figure 28.3*a*). In positive tubes, the cells settle out over a larger area, and the edges are irregular (figure 28.3*b*). Record your results in table 28.3 of the Laboratory Report.

6. Examine the tubes *quantitatively* for agglutination by gently mixing the contents (easily accomplished by flicking the tube back and forth with your index finger), and observing the tube with a good light source against a dark background. Record the amount of agglutination in table 28.3 of the Laboratory Report as follows:

   – no agglutination, cells remain in a cloudy suspension
   1 + approximately 25% cell clumping, supernatant cloudy
   2 + approximately 50% cell clumping, supernatant moderately cloudy
   3 + approximately 75% cell clumping, supernatant slightly cloudy
   4 + 100% cell clumping, supernatant clear

Figure 28.3 Examination of tubes in an agglutination test. (*a*) Negative control and (*b*) positive agglutination result.

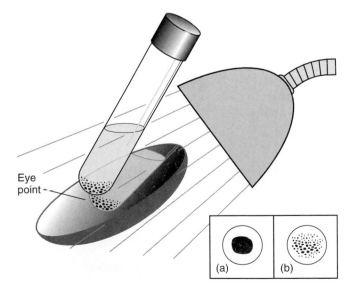

Eye point

(a) (b)

The Use of a Nonmicrobial Antigen (Cardiolipin) for Initial Syphilis Diagnosis (the USR Test)

Note: The USR test is performed as described in the *USPHS Manual of Tests for Syphilis 1969* and its supplement, January 1982.

1. Thoroughly soak and wash two glass slides in a glassware detergent solution. Then rinse with tap water 3–4 times, followed by a distilled water rinse, finally wiping dry with a clean lint-free cloth. This cleaning procedure will enable the serum to spread evenly within the inner surface of the circle.

2. With a wax-marking pencil, inscribe two circles (14-mm diameter) on each of the two slides.

3. The syringe with needle should be washed by prerinsing with tap water, then soak and wash thoroughly in a glassware detergent solution. Follow by rinsing with tap water 6–8 times, then with distilled water, absolute alcohol, and acetone, respectively. Finally, air dry until the acetone odor is gone.

4. The syringe with needle attached should be calibrated for delivery by filling the syringe with antigen suspension. Then hold the syringe in a vertical position and expel the suspension dropwise into the suspension bottle. Count the number of drops delivered per ml of antigen. The needle should deliver 45 drops ± 1.* Adjust drops per ml by either narrowing the open end of the needle to allow more drops per ml to be delivered or opening the end of the needle to allow fewer drops per ml.

5. Using a rubber bulb, pipette 0.05 ml of positive antiserum in the center of one of the circles using a 0.2-ml serological pipet. Spread the serum with the aid of the pipet tip over the area of the circle. Discard the used pipet in a hazardous waste container.

6. To each circle, add 1 drop of cardiolipin antigen using the previously calibrated syringe.

7. Rotate the slide for 4 minutes, preferably on a rotating machine at 180 rpm, circumscribing a circle ¾″ in diameter on a horizontal plane. The slide may also be rotated by hand for 4 minutes (see figure 28.1).

*Note: If unable to properly calibrate the syringe, use a 50 microliter (µl) pipet to deliver 0.05 ml of antiserum and antigen.

8. Observe the slide *immediately* with the 10× objective of your microscope and record your findings in part 3 of the Results as Reactive, Weakly Reactive, or Nonreactive, as determined by the amount of clumping:
   Reactive = medium and large clumps
   Weakly reactive = small clumps
   Nonreactive = no clumps or slight roughening
9. Repeat the procedure with the negative control antiserum, and then again with the weakly reactive antiserum.

Note: Each serum reported as positive in the USR qualitative test would normally be subject to further serologic study, including quantitation using a method somewhat like the rapid slide test, and if indicated, to other confirmatory syphilis serology tests, for example, the Fluorescent Treponemal Antibody Absorption (FTA-ABS) or the Hemagglutination Treponemal Test (HATTS). Thus, positive tests with the USR antigen are not conclusive evidence for syphilis. Conversely, a negative USR test by itself does not rule out syphilis diagnosis.

NOTES:

Name _____ Date _____ Section _____

Laboratory Report: Traditional Agglutination Reactions Employing Microbial and Nonmicrobial Antigens

Results

1. Rapid slide test with a microbial antigen
   a. Identify the *Proteus* antigen you used: _____
   b. Record the amount of agglutination in table 28.2 below: _____

Table 28.2 Rapid Slide Procedure: A Qualitative Test for Measuring *Proteus* Agglutination with Different Serum Dilutions

Dilution:	1:20	1:40	1:80	1:160	1:320	Saline control

Amount of agglutination:

   c. Record the highest serum dilution with 2+ agglutination: _____
   d. Record the serum titer as the reciprocal of the highest dilution showing a 2+ reaction: _____
2. Tube dilution test with microbial antigen
   a. Identify the *Proteus* antigen you used: _____
   b. Record the amount of agglutination for the various dilutions in table 28.3 below:

Table 28.3 Tube Dilution Test: A Quantitative Test for Measuring *Proteus* Agglutination with Different Serum Dilutions

Dilution:	1:20	1:40	1:80	1:160	1:320	1:640	1:1280	1:2560	1:5120	Control

Amount of Agglutination:

   c. Record the highest serum dilution with 2 + agglutination: _____
   d. Record the serum titer as the reciprocal of the highest dilution showing a 2+ reaction: _____
3. Reagin test for syphilis with a nonmicrobial antigen (cardiolipin)

   a. Record your results for the positive, weakly positive, and negative control serums:

   Positive serum:_____ Weakly positive serum:_____ Negative serum:_____

Questions

1. What titers did you find in the bacterial agglutination tests?

_____Rapid slide   _____Tube dilution

If this were serum from a patient, what would be its significance with respect to the probability of infection?

2. Why is a positive titer with the Weil-Felix test not necessarily conclusive that the patient has a rickettsial infection?

3. What other tests would be necessary to confirm a rickettsial infection? Consult your text and other references for possible answers, for example, the *Difco Manual*.

4. The rapid slide agglutination test is defined as a qualitative method. Why then are a series of doubling dilutions evaluated for use with this method?

5. Discuss ways in which the rapid slide agglutination test differs from the USR agglutination test. Consider variables such as test antigens and observation techniques.

6. Discuss the pros and cons of using nontreponemal antigens (such as cardiolipin) rather than treponemal antigens for syphilis diagnosis.

NOTES:

EXERCISE

Lancefield Grouping of Pathogenic Streptococci with a Latex Slide Agglutination Test

Getting Started

The genus *Streptococcus* deserves special attention because of its involvement in numerous diseases of humans and animals (see exercises 22 and 23.) It represents the predominant normal bacterial flora of the human respiratory tract, and is also found in the intestinal and genital tracts. Few other microorganisms of medical importance can elaborate as many **exotoxins** and destructive enzymes, as well as produce serious infections in virtually every tissue as can *streptococci* (table 29.1). One of the **sequela** diseases, rheumatic fever, was recently reported on the upsurge in the United States after decades of steady decline (Bisno, 1988). It is caused by a Lancefield Group A streptococcus that initially causes pharyngitis (strep throat).

It is important to differentially diagnose strep throat infections from viral throat infections since both acute rheumatic fever and acute **glomerulonephritis** can occur if prompt appropriate therapy is not initiated (Facklam and Carey, 1985). Rapid immunological slide tests are now available for detecting and differentiating pathogenic streptococci. They have their origin in historical research done by Rebecca Lancefield (1933), who was able to group streptococci serologically into 18 groups (A through R) on the basis of a group specific carbohydrate antigen present in their cell wall. One exception is the Group D streptococci in which a noncarbohydrate antigen, teichoic acid, is found in the cytoplasmic membrane. The serological test developed by Lancefield was an overnight capillary **precipitin** test using extracted cell wall antigens.

Further examination of table 29.1 shows that most human streptococcal diseases are caused by beta (β)-hemolytic streptococci (see exercise 23 for how to determine **β-hemolysis**) belonging to Lancefield Group A. This species, *S. pyogenes*, in contrast to other pathogenic streptococci, is found in only a small percentage of healthy humans. Other characteristics important for final identification of streptococci include biochemical and morphological tests (see exercises 22 and 23). In conjunction with this exercise, you examine their morphology and hemolysis reactions on blood agar. With one exception, *S. pneumoniae*, all streptococci appear as chains when grown in a broth medium (figure 29.1*a*). On an agar medium they sometimes appear as diplococci (figure 29.1*b*). *S. pneumoniae* is a true diplococcus that can be differentiated from other diplococci by its lancet shape (figure 29.1*c*) and **encapsulation.**

Table 29.1   Some Characteristics Useful in Identifying the Major Streptococci Associated with Human Disease

Streptococcus Species	Lancefield Group	Hemolysis Type	Human Diseases
S. pyogenes	A	β	Pharyngitis, scarlet fever, wound infections, skin, ear, lungs, other tissues, sequela: glomerulonephritis, rheumatic fever
S. agalactae	B	β	During birth and in infants: may develop pneumonia, septicemia, meningitis
S. pneumoniae	B	α	Bacterial pneumonia, ear infections, meningitis
S. faecalis	D	α or β	Subacute bacterial endocarditis (rare),
S. faecium			urinary tract infections
S. durans			
S. viridans			
(10 species)	Occasional false +	α	Subacute bacterial endocarditis

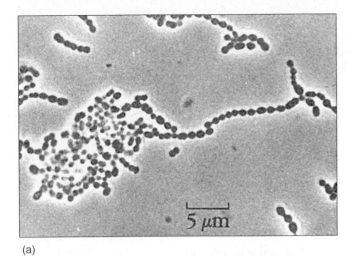

(a)

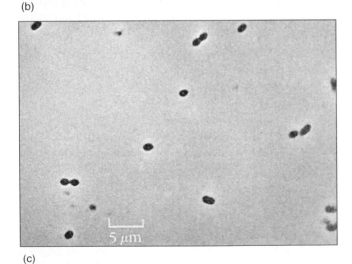

(b)

(c)

Figure 29.1 Genus *Streptococcus* morphology.
(a) *S. pyogenes* chains in a broth medium;
(b) *S. pyogenes*, sometimes diplococci on solid media;
(c) *S. pneumoniae*, lancet-shaped diplococci in a broth medium. Phase-contrast photomicrographs courtesy of G. E. Roberts.

The principle of the latex agglutination test for Lancefield grouping of pathogenic streptococci is that of the streptococcus cell wall carbohydrate antigens being allowed to react with specific antibodies coated to the surface of latex beads. The agglutination reaction occurs instantaneously, producing a latex particle aggregation large enough to be seen visually with the naked eye, thus eliminating the need for microscopic observation. A necessary preliminary step is the extraction of specific cell wall carbohydrate antigens. The carbohydrate known as "C" carbohydrate is a branched polymer composed of N-acetylglucosamine coupled with certain sugar molecules. It can be extracted from the cell wall by either hot formamide or hot trichloroacetic acid autoclaving, or by enzymatic digestion with lysozyme (see exercise 27) and certain microbial enzymes. The enzymatic process is used in this exercise.

Using the previously discussed identification methods (hemolysis, morphology, and latex agglutination), you have an opportunity in this exercise to differentiate two streptococcus species: a Lancefield Group A species (*S. pyogenes*) and a Lancefield Group B species (*S. pneumoniae*).

Definitions

Beta-hemolysis. A sharply defined, clear, colorless zone of hemolysis surrounding colonies of certain streptococci grown on blood agar plates.

Encapsulation. The surrounding of some bacteria by a protective gelatinous material, which may also relate to their virulence.

Exotoxin. A soluble, poisonous protein that passes into the growth medium during the growth of certain bacteria.

Glomerulonephritis. Inflammation of the kidneys affecting the structure of the renal glomeruli (inflammation of the capillaries caused by toxins produced elsewhere in the body).

Precipitin reaction. The reaction of an antibody with a soluble antigen to form an insoluble, visible antigen-antibody complex.

Sequela. An aftereffect of disease or injury that is often more serious than the initial disease.

Objectives

1. To provide introductory information about the medical importance of the genus *Streptococcus* and how to identify them by a combination of morphological, biochemical (hemolysis), and immunological (latex agglutination) tests.
2. To determine the Lancefield Group, A or B, of two *Streptococcus* species with the latex slide agglutination test.

References

Biano, Alan, U. of Miami School of Medicine, in the *Seattle Times*, p. A4, January 20, 1988.

Facklam, R. R. 1980. "Isolation and identification of streptococci." HEW Publication, U.S. Dept. of Health, Education, and Welfare, Centers for Disease Control, Atlanta.

Facklam, R. R., and Carey, R. B. 1985. "Streptococci and aerococci," 154–175. In E. H. Lennette, A. Balows, W. J. Hausler, Jr., and H. J. Shadomy (ed.), *Manual of clinical microbiology*, 4th ed. Washington, D.C.: American Society for Microbiology.

Lancefield, R. C. 1933. Serological differentiation of human and other groups of hemolytic streptococci. *J. Exp. Med.* 57:571–595. A classical paper worth examining.

Nester et al. *Microbiology: A human perspective*, 4th ed., 2004. Chapter 17, Section 17.4; Chapter 22, Section 22.3; and Chapter 23, Section 23.3.

Slifkin, M., and Pouchet-Melvin, G. R. 1980. "Evaluation of three commercially available test products for serogrouping beta-hemolytic streptococci." *Journal of Clinical Microbiology* 11:249–255.

Materials

Fresh (24 hr, 37° C) *unknown* cultures of *S. pyogenes* (Lancefield Group A) and *S. pneumoniae* (Lancefield Group B) on blood agar plates labeled 1 and 2

Fresh TS broth cultures of unknowns 1 and 2 above

Diluted cell wall extraction enzymes 0.3-ml aliquots contained in 3 serological test tubes

One vial with dropper of a 1% latex bead suspension coated with strep Gr A antibodies, prepared in a glycine buffer (to be shared by class)

A vial (similar to vial above) except coated with strep Gr B antibodies (to be shared by class)

A vial of polyvalent positive control antigen (an extract of strep Gr A, B, C, F, and G). To be shared by class.

Sterile physiological saline (0.85% NaCl)

Calibrated 1-ml pipets, 2

Toothpicks

Disposable plastic droppers

37°C water bath

Vortex for mixing tubes

High-intensity incandescent light source

Mechanical rotator for slides (if available)

Ruler calibrated in mm

Procedure

Safety Precautions: Because β-hemolytic streptococci are opportunistic pathogens (see exercise 23), all used slides, disposable pipets, and stirring sticks should be disposed of in the hazardous waste container. Any used glassware (blood agar plates, broth cultures, and tubes containing extraction enzyme-unknown bacteria) should be promptly autoclaved. If any culture material is spilled, notify your instructor.

1. Prepare Gram stains of both the broth and agar unknown cultures. Examine with the oil immersion objective and make drawings in the circled areas of the Laboratory Report. Look for differences in form, shape, and size.
2. If the Gram stains are indicative of streptococci (Gram-positive cocci in pairs or chains), note their hemolysis pattern on blood agar, enter in the Laboratory Report, and proceed as follows:
   a. With a 1-ml pipet, transfer 0.3 ml of extraction enzyme to each of three small

test tubes. Label the tubes: 1 for unknown #1, 2 for unknown #2, and 3 for negative control.

b. With a sterile loop, remove a single sweep of confluent growth from unknown blood agar culture #1 and transfer the contents to the #1 tube of extraction enzyme. Vortex the tube, then incubate it in 37°C water bath for 30 minutes.

c. Repeat step b with unknown #2.

d. Incubate tube #3, the negative control, in the water bath.

e. Following incubation, add 0.3 ml of sterile physiological saline to each tube. Mix well by vortexing.

f. Thoroughly mix the two vials of antibody-coated latex bead suspensions (A and B) by manual shaking. Make sure the beads are entirely resuspended.

g. With a wax-marking pencil, inscribe two 14-mm circles on each of four clean glass slides. Label the slides as shown in figure 29.2.

h. Dispense one drop of Lancefield Group A latex bead suspension near the center of each of the four circles on slides 1 and 3 (see figure 29.2).

i. Repeat step h for Lancefield Group B latex bead suspension on slides 2 and 4.

j. With a dropper, transfer 1 drop of unknown #1 organism-enzyme extract suspension near the center of the left circle on slides 1 and 2. Discard dropper in the hazardous waste container.

k. Repeat step j with unknown #2 organism-enzyme extract suspension on the right circle of slides 1 and 2.

l. Add 1 drop of the polyvalent positive control reagent near the center of the left circle on slides 3 and 4.

Figure 29.2 Procedural arrangement for addition of antigens and antibodies to latex agglutination slides.

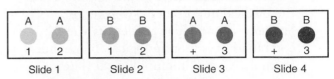

A = Lancefield Group A antibody-coated latex bead suspension
B = Lancefield Group B antibody-coated latex bead suspension
1 = Unknown #1 organism-enzyme extract
2 = Unknown #2 organism-enzyme extract
3 = Negative control tube #3
+ = Polyvalent positive antigen control

m. With a dropper, transfer 1 drop of the negative control extraction enzyme (tube 3) near the center of the right circles on slides 3 and 4 (see figure 29.2).

n. With a stirrer (such as a toothpick), mix the contents of each circle.

Note: Use a clean stirrer for each circle.

o. Rock each slide for two minutes either on a mechanical rotator (95–110 rpm) or gently by hand with a rocking motion (see figure 28.1).

p. Examine each freshly prepared slide for agglutination with a high-intensity incandescent light source.

Note: Delays in reading agglutination reactions can result in drying of slides, which could render the results uninterpretable.

q. Report your agglutination test results in table 29.2 of the Laboratory Report.

Note: In reading the slides, you should first compare the agglutination patterns of the positive and negative controls before proceeding to the unknowns.

EXERCISE

Laboratory Report: Lancefield Grouping of Pathogenic Streptococci with a Latex Slide Agglutination Test

Results

1. Drawings of unknown Gram-stained bacteria seen with the oil immersion objective:

    Unknown #1

    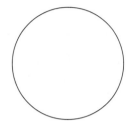   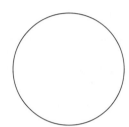

    Blood agar   TS broth

    Unknown #2

    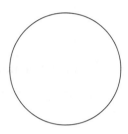   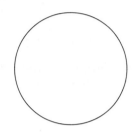

    Blood agar   TS broth

2. Describe the type of hemolysis found on blood agar:

    Unknown #1:

    Unknown #2:

3. Record the latex agglutination reactions (+ or −) in table 29.2.

Table 29.2   Latex Agglutination Reactions with Lancefield Group A and Group B *Streptococcus* Antisera

Test Antigen	Lancefield Group A Antiserum	Lancefield Group B Antiserum
1. Unknown #1		
2. Unknown #2		
3. Negative control		
4. Polyvalent + control		

4. From the three studies (morphology, hemolysis, and latex agglutination), which unknown did you identify as *S. pyogenes* _____

   and as *S. pneumoniae* _____?

5. Were all of your findings consistent with the literature? If not, describe any inconsistencies observed and if possible provide an explanation.

Questions

1. Why are Group D streptococci not included in the polyvalent positive control?

2. What are some reasons for including positive and negative controls?

3. What advantages are there of the latex slide test over the capillary precipitin test for Lancefield grouping of pathogenic streptococci?

4. *Streptococcus pneumoniae* bacteria often possess cell wall surface antigens that react with Lancefield Group C antiserum (Slifkin and Pouchet-Melvin, 1980). In this event, how would you determine if the positive agglutination result is due to a Group C *Streptococcus* or to *S. pneumoniae?*

NOTES:

EXERCISE

30

Use of an Enzyme-Linked Immunosorbent Assay (ELISA) Test for *Coccidioides immitis* Identification

Getting Started

The ELISA test is widely used for identification of both plant and animal pathogens, including viruses. It is also used as an initial screening test for detecting antibodies to the AIDS virus. In the clinical setting, it is used for identification of a variety of microbial pathogens because of test sensitivity and simplicity, often requiring only a swab sample from the infected host. It serves as an example of using a combined precipitin-enzyme reaction to achieve an end point. Of great importance to the success of this technique is the plastic microwell plate which can attract the reactants to its surface and hold on to them (see paragraph 2). In this exercise, the ELISA technique is used to identify a medically important dimorphic fungus, *Coccidioides immitis*, that assumes a yeastlike phase in the body of the host (see color plate 17). In culture it produces a typical mold colony containing barrel-shaped arthrospores (see color plates 15 and 16).

Various modifications of the ELISA antigen-antibody technique exist. The one used here is an indirect immunosorbent assay (figure 30.1a). (step 1) With this method, the patient's antiserum is added to a microwell previously coated with a mixture of *Coccidioides immitis* antigens. If antibodies related to the coccidioidal antigens are present, they become attached to the adsorbed antigens. (step 2) After washing to remove unbound specimen components, an antibody conjugate that has been coupled to the enzyme, horseradish peroxidase, is added. If binding occurs between the antigen and the antibody conjugate, a sandwich is formed containing adsorbed antigens, patient antibodies, and the horseradish peroxidase enzyme. Peroxidases are enzymes which catalyze the oxidation of organic substrates. (step 3) Next the organic substrate used for this test, urea peroxide, is added. When oxidized by the peroxidase enzyme, free oxygen (O) is released. A color indicator, tetramethylbenzidine, is added, which when oxidized by the free oxygen produces a yellow color. Lack of color means that the patient's antiserum does not contain *Coccidioides immitis* antibodies.

Another widely used ELISA method is the direct immunosorbent assay commonly known as the double antibody sandwich assay (see figure 30.1b). It is widely used with plants to determine if they are infected with various pathogenic bacteria and viruses. Some of the symptoms include spotting and discoloration of the leaves, decreased yield, etc. Some of these pathogens are able to infect a wide variety of plants. For example tomato spotted wilt virus and impatiens necrotic spot virus are able to infect a variety of flowers and vegetables. Examples include hydrangeas, begonias, orchids, potatoes, etc.

References

Kaufman, L., and Clark, M.J. 1974. "Value of the concomitant use of complement fixation and immunodiffusion tests in the diagnosis of coccidioidomycosis." *Applied Microbiol.* 28:641–643.

Nester et al., *Microbiology: A human perspective*, 4th ed., 2004. Chapter 17, Section 17.7.

Pappagianis, D., and Zimmer, B.L. 1990. "Serology of Coccidioidomycosis." *Clin. Microbiol. Reviews* 3:247–268.

Materials

A Meridian Diagnostics, Inc.* Premier *Coccidioides* EIA Package Insert that contains:

Antigen coated microwells (96)—breakaway plastic microwells, each coated with a mixture of TP (tube precipitin) and CF (complement fixation) antigens.

Positive control (2.7 ml)—prediluted positive human serum with a preservative. Do not dilute further.

Note: The positive control serum and all materials which they contact should be

*We wish to thank Meridian Diagnostics, Inc. for their help in providing for educational use Premier *Coccidioides* Enzyme Immunoassay test kits at a reasonable cost.

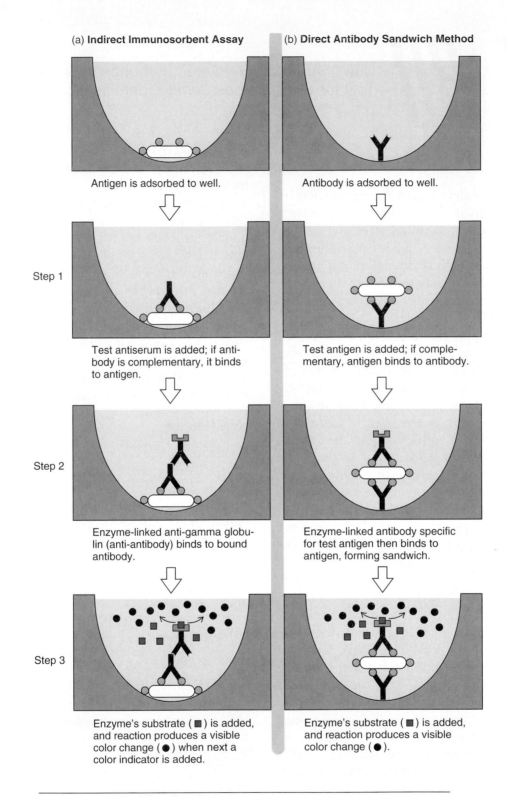

(a) Indirect Immunosorbent Assay

Antigen is adsorbed to well.

Step 1

Test antiserum is added; if antibody is complementary, it binds to antigen.

Step 2

Enzyme-linked anti-gamma globulin (anti-antibody) binds to bound antibody.

Step 3

Enzyme's substrate (■) is added, and reaction produces a visible color change (●) when next a color indicator is added.

(b) Direct Antibody Sandwich Method

Antibody is adsorbed to well.

Test antigen is added; if complementary, antigen binds to antibody.

Enzyme-linked antibody specific for test antigen then binds to antigen, forming sandwich.

Enzyme's substrate (■) is added, and reaction produces a visible color change (●).

Figure 30.1 The ELISA technique. (*a*) The indirect antibody method and (*b*) the double antibody sandwich method.

handled at Biosafety Level 2 as recommended in the CDC/NIH manual "Biosafety in Microbiology and Biomedical Laboratories," 1988. In view of using positive control serum, this exercise can only be conducted by personnel trained in handling pathogens. Laboratory instructors and students well grounded in proper use of aseptic technique should have no difficulty in meeting this requirement. The laboratory facilities must be at least Level 1, standard open bench, typical of most microbiology teaching laboratories.

Sample diluent (50 ml)—buffered protein solution with a preservative

20X wash buffer (50 ml)—concentrated wash buffer with a preservative

IgM enzyme conjugate (10 ml)—affinity purified goat anti-human IgM antibodies conjugated to horseradish peroxidase in buffered protein solution containing a preservative

IgG enzyme conjugate (10 ml)—affinity purified goat anti-human IgG antibodies conjugated to horseradish peroxidase in buffered protein solution containing a preservative

Horseradish peroxidase substrate (10 ml)—buffered solution containing urea peroxide and tetramethylbenzidine

Stop solution (10 ml)—2N sulfuric acid. **CAUTION:** Avoid contact with skin. Flush with water if contact occurs.

Microwell strip holder

Note: The expiration date is on the kit label. Store kit at 2°–8°C and return kit promptly to the refrigerator after use. Microwells not being used must be removed from the microwell holder and placed back inside the resealable foil pouch and sealed. It is important to protect the strips from moisture.

Other Materials

Pipets capable of delivering 10, 20, 100, and 200 ul

Test tubes (12 × 75 mm) for dilution of sample

Distilled or deionized water

Squirt bottle

Timer

If available a plate reader capable of reading absorbance at 450 nm. A dual wavelength reader is preferred, using a second filter of 630 nm to correct for light scatter.

Note: The plate reader is not necessary for reading positive control results. The definite yellow color is easily observed with the naked eye.

Reagent Preparation

1 Bring entire kit, including microwell pouch, to 22°–25°C before and during use. Warming requires at least 1 hour.

2 Prepare sufficient 1X wash buffer for use by measuring 1 part of 20X buffer and diluting with 19 parts of water. The 1X buffer can be stored at room temperature for up to one month. Discard if buffer becomes contaminated. For performing washes, the buffer can be transferred to a wash or "squirt" bottle.

Procedure

Note: Each microwell is coated with both TP (tube precipitin) and CF (complement fixation) antigens. It is recommended that both the IgM and the IgG antibody assays be performed simultaneously for the serum samples.

1. Snap off a sufficient number of microwells for positive serum sample and sample control and insert them into the microwell holder. Record sample positions, as shown in the example below.

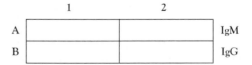

2. Add 100 ul of positive control antiserums, IgM and IgG, conjugated with horseradish peroxidase enzyme, to each of two separate microwells (A1 and B1) and 100 ul of sample diluent (SD) to each of two separate microwells (A2 and B2).

The latter two wells represent negative controls. Any remaining wells will not be used in this exercise because they are for use with patient sera.

3. Mix the samples by gently rotating the microwell for 10–15 seconds on the countertop.
4. Immediately following step 3 incubate the microwell at 22°–25°C for 30 minutes.
5. Hold the microwell plate firmly from the bottom and gently squeeze it.
   a. Dump plate contents into a biohazard receptacle, next strike the inverted plate firmly on a clean stack of paper towels or other absorbent material.
   b. With wash bottle, fill all wells with 1X wash buffer by directing the stream of buffer to the sides of the wells to prevent foaming. Dump the plate contents in the biohazard receptacle and strike the plate on the towels.
   c. Repeat step 5b two more times.

Note: The secret for success with this exercise lies in thorough rinsing of the microwells. After the final wash, strike plate on fresh towels hard enough to remove as much excess wash buffer as possible but do not allow wells to completely dry at any time.

6. Add two drops of IgM enzyme conjugate to the microwells of row A.
7. Add two drops of IgG enzyme conjugate to the microwells of row B. Mix the samples by gently rotating the microwell plate for 10–15 seconds on the countertop.
8. Immediately following step 7 incubate at 22°–25°C for 30 minutes.
9. Repeat the step 5 washing cycle.
10. Add two drops of the urea peroxide and tetramethylbenzidene substrate to each microwell. Start timer with addition of substrate to the first well. Mix by gently rotating the microwells 10–15 seconds on the countertop.
11. Incubate at 22°–25°C for 5 minutes.

12. Carefully add two drops of stop solution (2N sulfuric acid) to each microwell in the same order as step 10. Mix as in step 10 and wait 2 minutes before reading. A yellow color on the bottom of the control serum wells (A1 and B1) represents a positive test. The bottom of the negative control wells (A2 and B2) should appear clear. Readings should be made within 15 minutes. Enter your results in table 30.1 of the Laboratory Report.
13. If a plate reader is available you may be able to quantify your results. Carefully, wipe the underside of the microwells with a lint-free tissue and measure the absorbance at 450 nm as follows:
   a. Blank on air.
   b. Read the negative control microwells, values should be <0.100.
   c. Then reblank the reader using the negative control well values.
   d. Enter your results in table 30.2 of the Laboratory Report.
14. Disinfect and retain microwell holder. Discard used assay materials in biohazard container.

Note: The positive and negative controls must be assayed with each batch of patient specimens to provide quality assurance. The negative control (sample diluent) should yield an A 450 value <0.100 with both IgM and IgG conjugates when blanked on air. The positive control microwells should have a definite yellow color and yield an A 450 value >0.500 and <2.500 with both IgM and IgG conjugates.

For patient specimens the results have been interpreted as follows:

Negative = Absorbance Value <0.150
Indeterminate = Absorbance Value >0.50 but <0.199
Positive = Absorbance Value >0.200

Interpretation of Results

A negative result with both IgM and IgG indicates that serum antibody to C. *immitis* antigens is either absent, below the level of detection of the assay, or the specimen was obtained too early in the response. A positive result with either IgM or IgG implies the presence of antibody to C. *immitis*. A positive response with either conjugate should be reported. An early acute phase patient may only present an IgM response, while the chronic or convalescent patient may only present an IgG response.

Specimens that yield an indeterminate result should be retested. If the retest result is still indeterminate, a second specimen should be obtained. Extremely strong positive reactions may yield a purple precipitate. Absorbances obtained with such reactions may be lower than expected but will still be positive.

Limitations of the Procedure

A negative result with both IgM and IgG antibodies does not preclude diagnosis of coccidioidomycosis, particularly if only a single specimen has been tested and the patient shows symptoms consistent with a positive diagnosis.

A positive ELISA result should be confirmed by an Ouchterlony immunodiffusion assay (see exercise 31). However because of the relative insensitivity of the ID (immunodiffusion) procedures, an ID negative test does not prevent the possibility of coccidioidomycosis (see Kaufman and Clark, 1974). Diagnosis is based on both laboratory and clinical findings as well as the presence of antibody.

Positive results with either IgM or IgG (but not both) also suggest coccidioidal disease, but in different disease states. An early acute phase patient may only present an IgM response, while the chronic or convalescent patient may only present an IgG response.

Such results should be compared with patient symptoms to determine if there is a logical correlation.

NOTES:

EXERCISE

30

Laboratory Report: Use of an Enzyme-Linked Immunosorbent Assay (ELISA) Test for *Coccidioides immitis* Identification

Questions

1. Discuss the test results and their significance.

Table 30.1 Visual Observation of Processed Samples

Sample	Yellow Color (I)
IgM	
IgM control	
IgG	
IgG control	

Table 30.2 Sample Absorbance at 450 nm with a Plate Reader

Sample	Absorbance
Negative Control	
IgM	
IgG	

2. Discuss laboratory safety considerations related to handling of:

a. Antigen-coated microwells

b. Positive serum control

c. Immunoglobulin enzyme conjugates

d. Urea peroxide

e. Tetramethylbenzidine

3. What is the importance of rinsing when conducting the ELISA test?

4. Why is the ELISA test for coccidioidomycosis a more definitive test than the Ouchterlony immunodiffusion test (exercise 31)? Is the latter test still of value as a diagnostic tool? Explain your answer.

5. Discuss the pros and cons of using the enzyme-linked immunosorbent assay (ELISA) as used here and the double antibody sandwich ELISA assay. You may need to consult your text for the answer to this question.

EXERCISE

31

An Ouchterlony Double Immunodiffusion Test for *Coccidioides immitis* Identification

Getting Started

Precipitin reaction tests such as the Ouchterlony test are widely used for **serodiagnosis** of fungal diseases (Kaufman and Reiss, 1985). More recently the ELISA test (see exercise 30) has also been used for this purpose (de Repentigny and Reiss, 1984). It is not unusual to use two or more serological tests for initial diagnosis of a fungal disease (Rippon, 1974, p. 382). The latex slide agglutination test (see exercise 29) has also been used as an initial diagnostic test for fungal disease.

With the Ouchterlony procedure, soluble antigen and **serum** solutions containing antibodies are placed in separate wells of an agar base. The reactants diffuse from the wells and form thin, white precipitin line(s) where they meet in optimum proportions (figure 31.1*a*). Since both immunoreactants move in this system, it is known as a **double immunodiffusion test.** In a related system, the Oudin technique, diffusion occurs only in a single dimension (figure 31.1*b*).

According to the **lattice theory** depicted pictorially in figure 31.2, the precipitin lines form best where an excess of antibody relative to antigen is present. When the Ab/Ag ratio is less than 1, soluble complexes rather than precipitates occur (figure 31.2*d–f*). In some instances, more than one precipitin line will form in the agar, because the antigen preparation contains more than one type of antigen; thus, each band formed represents one antigen-antibody reaction.

The curvature of the precipitin line can provide information about the molecular weight of the antigen, providing the antigen and antibody are present in nearly equal amounts. The precipitin line appears straight if the antigen and antibody have about the same molecular weight. If the antigen has a higher molecular weight, the line is concave toward the antibody well; if the antigen is of lower molecular weight, the line is concave toward the antigen well. These relationships are derived

Figure 31.1 Some commonly used systems for gel diffusion precipitin reactions. (*a*) Double diffusion in two dimensions (Ouchterlony technique) in which diffusible antigen and antibody solution are placed in separate wells cut in an agar plate. Direction of diffusion is shown by (arrows) (*b*) Single diffusion in one dimension (Oudin technique) using a soluble diffusion antigen layered over an antibody contained in an agar gel column. Black lines are opaqued precipitin bands. Reproduced with the permission of Meridian Diagnostics, Inc., Cincinnati, Ohio 45244.

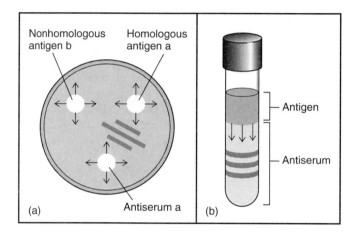

from the principle that the rates of diffusion of any mixture increase with concentration and decrease with molecular weight.

Finally, antigens or antibodies can be tested for **identity** by placing a test well of the substance in question adjacent to the wells of a known system. If the antigen-antibody complexes are identical, the precipitin lines form an unbroken line of identity with the known system (figure 31.3*a*). **Partial identity** and **nonidentity** reactions are also possible (figure 31.3*b* and *c*, respectively). A partial identity reaction occurs when certain components of the antigens (or antibodies) are identical and others are not. The **spur** represents the components that are unrelated. A nonidentity reaction occurs when the antigen-antibody complexes are different. The resulting "X," or cross reaction, indicates that two unrelated complexes are present.

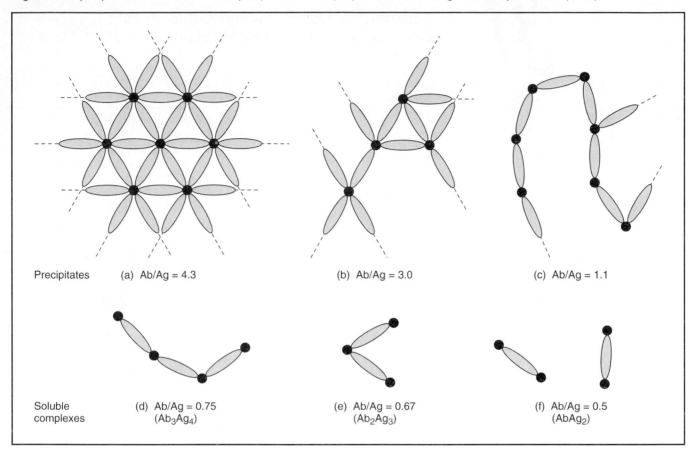

Precipitates   (a) Ab/Ag = 4.3   (b) Ab/Ag = 3.0   (c) Ab/Ag = 1.1

Soluble   (d) Ab/Ag = 0.75   (e) Ab/Ag = 0.67   (f) Ab/Ag = 0.5
complexes   (Ab_3Ag_4)   (Ab_2Ag_3)   ($AbAg_2$)

The test organism chosen to illustrate principles of Ouchterlony double immunodiffusion is a sometimes pathogenic mold, *Coccidioides immitis*. In nature it survives best in desert soils where the temperatures average 100°F in the summer and in the fall and winter 33° to 38°F. It actually grows better on rich soil than on poor. However, survival is greatly reduced on such a soil containing normal bacteria and fungal flora. Thus the most highly endemic regions for contracting coccidioidomycosis in North America are the southwestern United States and northern Mexico. When first examined in 1892 in lesions of patients it was likened to protozoa in the order Coccidia. Later studies showed it was a mold (see color plates 15, 16, and 17). At first the morphology of the endosporulating spherule, when examining tissue, suggested a relation to the protozoa. Later laboratory studies establishing the same endospore formation within spherules implied a relationship to

the fungal class Zygomycetes (see Rippon, 1974). We now know that it is a dimorphic yeastlike fungus belonging to the class Ascomycetes. Nevertheless the genus name *Coccidioides* remained. The species name *immitis* means (*im* = not, *mitis* = mild). Some common synonyms for this disease are San Joaquin Valley fever, Valley bumps (small tender reddened nodules under the skin), and California disease.

In this study rather than using a live fungus for an antigen source, a purified culture filtrate of *Coccidioides immitis* containing the "F" antigen at a concentration of 100 units/ml is used. The antiserum source was produced in hyperimmune (not infected) goats injected with the purified "F" antigen.

An interesting alternative method for study of the Ouchterlony procedure is one which uses various unknown meat samples as an antigen source and for antisera (antibodies) anti-horse, pig, and cow albumin samples.

Figure 31.3 Double diffusion precipitin reactions observed in agar gel plates:
1. Antigen solutions with one or more antigenic components. Antigen solution A contains two distinct antigen components as indicated by precipitin lines a_1 and a_2.
2. Lines of identity (fusion). (*a*) Antigen A solution contains a component antigenically identical with antigen solution B. For example, A shows a line of identity by fusion and thickening of precipitin lines a_2 with b_2.
3. Lines of partial identity (spur). (*b*) Antigen C contains a component (c_2) that shares at least one antigenic determinant with antigen A, but A contains at least one antigenic determinant that is nonidentical to C. The evidence for the partial identity of antigens A and C is the merging and thickening of the C precipitin line with the A precipitin line, but the A precipitin line has a spur that overlaps the C precipitin line.
4. Lines of nonidentity (cross). (*c*) In this instance, the precipitin lines do not fuse but cross because the antigens A and D are not related chemically to one another.

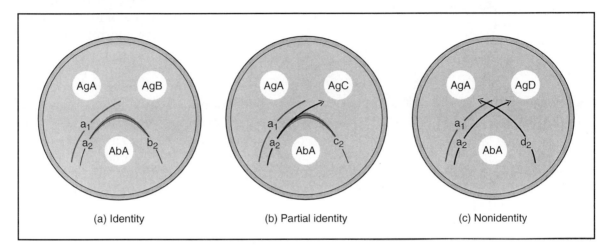

(a) Identity   (b) Partial identity   (c) Nonidentity

Definitions

Double immunodiffusion test. An immunological test in which both the antigen and the antibody move in the reaction system.

Immune response. Specific response to a foreign antigen characterized by the production of humoral antibodies or immune cells.

Lattice theory. A theory based on a framework formed by crossing soluble antigens and antibodies in a diagonal manner such that when present in the correct ratio of Ab to Ag, an aggregate forms, which when sufficiently large, precipitates out of solution in the form of a visible complex.

Precipitin reaction. The reaction of an antibody with a soluble antigen to form an insoluble precipitate.

Precipitin reaction test. A test in which an antibody is able to combine with an identifiable soluble antigen and cause aggregation and eventual precipitation out of solution.

Serodiagnosis. A diagnostic test employing the serum portion of the blood to test for antigen-antibody reactions.

Serum. The fluid portion of the blood that remains after the blood clots. Contains antibodies, hormones, and dissolved nutrients but not cells.

Objectives

1. To familiarize you with some of the principles of Ouchterlony double immunodiffusion, and to provide an opportunity for you to conduct and observe some of these principles at work in the laboratory.
2. To perhaps suggest some improvements in methodology once you have conducted this exercise.

References

Kaufman, L., and Reiss, E. 1985. "Serodiagnosis of fungal diseases," 924–944. In E. H. Lennette, A. Balows, W. J. Hausler, Jr., and H. Jean Shadomy, eds., *Manual of clinical microbiology*, 4th ed. Washington, D.C.: American Society for Microbiology.

Nester et al. *Microbiology: A human perspective*, 4th ed., 2004. Chapter 17, Section 17.4.

de Repentigny, L., and Reiss, E. 1984. "Current trends in immunodiagnosis of candidiasis and aspergillosis." *Review Infectious Diseases* 6:301–312.

Rippon, J. W. *Medical Mycology: The Pathogenic Fungi and Pathogenic Actinomycetes*. Philadelphia: W. B. Saunders, 1974.

Materials

A Meridian Diagnostics, Inc. Test Kit (603096),* which contains:

0.5 ml *Coccidioides* ID antigen

Diluted anti-*Coccidioides* ID control serum

Immunodiffusion agar plates

Capillary pipets with bulb

Moist chamber: a dish with a tight-fitting cover containing moist paper toweling is satisfactory provided the ID plates remain level and hydrated during the incubation period.

An inexpensive reading light is a desk lamp with a black cover which contains a 60-watt lightbulb for illumination. By holding the Ouchterlony plate vertically near the lower outside edge of the black cover you should be able to see well-defined precipitin lines.

Reagent quality or distilled water

*We wish to thank Meridian Diagnostics, Inc., for their willingness to provide short dated usable immunodiffusion kits at a reasonable cost.

Procedure

Note: Refer to figure 31.4 of the Laboratory Report for a description and numerical designation of the test well pattern of the ID agar plate.

1. Using a capillary pipet with an attached bulb, fill the pipet approximately ¾ full with *Coccidioides* antigen.

2. Next fill the center well (#7 on figure 31.4) with *Coccidioides* antigen.

   Note: Take care to avoid overflow of antigen solution from the well.

3. Using a fresh capillary pipet, repeat steps 1 and 2 with positive control antiserum. Fill well #1 with positive control serum. If you wish to observe a line of identity, also fill well 2 or 6 with positive control serum (see color plate 26).

4. If negative control sera or positive sera from patients known to have a *Coccidioides* infection are available, they can be added to other external wells of the ID agar plate. For example, if positive serum is available from a patient it should be placed in a well adjacent to one containing positive control antiserum. Such an arrangement will enable you to determine lines of identity, partial identity, or nonidentity (see figure 31.3 and color plate 26).

5. Identify the contents of each well in table 31.1 of the Laboratory Report. Also make a note in which wells, if any, the antiserum was diluted, and the amount of the dilution.

6. Place your name or initials and date on the ID plate cover. Incubate it in the moist chamber at room temperature for 24 to 48 hours.

7. After 24 to 48 hours incubation, read and record the ID bands in table 31.1 of the Laboratory Report. A light source is preferred for observing the nature of the bands (see Materials section). Particular attention should be paid to the orientation of the bands in relation to control serum bands. A smooth junction of the bands is indicative of an identity reaction (see color plate 26). If antiserum from a patient were included in this test, you would also look to see if the control band was bent toward a position in front of the patient well. This would indicate patient antibody at a low titer.

8. Interpretation of the test
   a. A band of identity with a known positive control indicates the presence of patient antibody against the antigen in question. In general, an identity reaction against a given antigen is indicative of active or recent past infection.
   b. Partial identity reactions are regarded as positive for antibody against the antigen only if no other identity reaction is present on the plate. Partial identity reactions are also indicators of possible disease.
   c. Nonidentity reactions are regarded as a negative test, and may also be apparent when the disease state is caused by a mycotic agent other than the one tested.
   d. The greater the number of precipitin lines observed, the greater the likelihood of severe disease. In arriving at a diagnosis, the test results should be interpreted together with all other lab and clinical data, including treatment history.
9. Limitations of the test. The high rate of negative serologic tests observed among culturally demonstrable cases limits the predictive value of a negative test. Positive test results must be confirmed culturally.

NOTES:

Name ______________________________ *Date* ______________ *Section* ______________

EXERCISE 31

Laboratory Report:
An Ouchterlony Double Immunodiffusion Test for *Coccidioides immitis* Identification

Results

Date ID plate incubated ___ Date observed___

1. Make drawings on figure 31.4 of all observed precipitin lines.

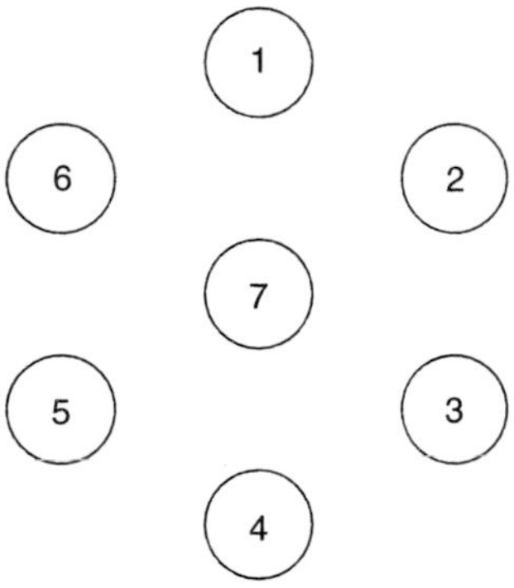

Figure 31.4 Test well pattern of the immunodiffusion agar plate.

2. Fill in the necessary blanks in table 31.1.

Table 31.1 Ouchterlony Fungal Immunodiffusion Analysis Form

Well No.	Well Description	Reading Observed
1	Positive control serum	
2		
3		
4		
5		
6		
7	*Coccidioides* antigen:	

Note: Indicate under "Well Description" in which wells, if any, the serum was diluted.

Note: Test is invalid if positive control is negative after 24-hour incubation.

3. Discuss your results and their significance.

4. What future studies might be suggested from your results?

Questions

1. Interpret the following *Coccidioides* immunodiffusion bands with respect to the possibility of having an active *Coccidioides* infection:

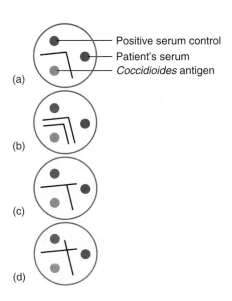

(a) — Positive serum control
— Patient's serum
— *Coccidioides* antigen

(b)

(c)

(d)

31–8 *Exercise 31 An Ouchterlony Double Immunodiffusion Test for* Coccidioides immitis *Identification*

2. Interpret the relative antigen (Ag) and antibody (Ab) molecular weights from the following immunodiffusion bands:

   a. ID test using equal concentrations of Ag and Ab.

      o – – Ab

      ———

      o – – Ag

   b. ID test using a greater concentration of Ag than Ab.

      o – – Ab

      ———

      o – – Ag

   c. ID test using equal concentrations of Ag and Ab.

      o – – Ag

      o – – Ab

3. Circle the Ag (•) Ab (⬬) diagram in which the best precipitin line is possible:

4. What advantages does Ouchterlony double immunodiffusion precipitin analysis have over Oudin single immunodiffusion analysis?

NOTES:

INTRODUCTION to the Prevention and Control of Communicable Diseases

Communicable, or infectious, diseases are transmitted from one person to another. Transmission is either by direct contact with a previously infected person, for example by sneezing, or by indirect contact with a previously infected person who has contaminated the surrounding environment.

A classic example of indirect contact transmission is an epidemic of cholera that occurred in 1854 in London. During a 10-day period, more than 500 people became ill with cholera and subsequently died. As the epidemic continued, John Snow and John York began a study of the area and were able to prove by **epidemiological methods** only (the bacteriological nature of illness was not known at that time) that the outbreak stemmed from a community well on Broad Street known as the Broad Street Pump* (figure I.10.1).

Then they discovered that sewage from the cesspool of a nearby home was the pollution source, and that an undiagnosed intestinal disorder had occurred in the home shortly before the cholera outbreak. They also learned that neighboring people who abstained from drinking pump water remained well, whereas many of those who drank pump water succumbed to cholera.

Today the incidence of cholera, typhoid fever, and other infectious diseases rarely reach **epidemic** proportions in those countries that have developed standards and regulations for control of environmental reservoirs of infection. The major reservoirs are water, food, and sewage.

The importance of epidemiology in tracing the source of an infectious disease is demonstrated in laboratory exercise 32. The exercise involves the use of a method for detecting a specific *Staphylococcus* strain on various body parts of the student class. If all members of a class carry the same strain, an epidemic is likely to exist.

The subject of public health sanitation is presented in exercise 33 (water microbiology).

*Snow, John: "The Broad Street Pump," in Roueche, Berton (ed.): *Curiosities of medicine*. Berkely, Medallion, ed., New York, 1964.

All of such studies come under the surveillance of public health agencies responsible for prevention and control of communicable diseases. Among these are the Department of Health and Human Services at the federal level, which has cabinet status, conducts preventive medicine research, provides hospital facilities for service men and women, and gives financial assistance to state and local health departments, as well as assistance at times to developing countries. Additionally, the Centers for Disease Control (CDC) in Atlanta play an important role in the prevention and control of communicable disease. Also, all states and other local government agencies perform important public health services.

Perhaps the most important international health organization is the World Health Organization (WHO) headquartered in Geneva, Switzerland. WHO distributes technical information, standardizes drugs, and develops international regulations important for the control and eradication of epidemic diseases. For example, smallpox, which was once a widespread disease, is virtually nonexistent today.

Finally, there are voluntary health organizations that help in some of the causes previously mentioned .

Definitions

Epidemic.   The occurrence in a community or region of a group of illnesses of similar nature, clearly in excess of normal expectancy.

Epidemiological methods.   Methods concerned with the extent and types of illnesses and injuries in groups of people, and with the factors that influence their distribution. This implies that disease is not randomly distributed throughout a population, but rather that subgroups differ in the frequency of different diseases.

Figure I.10.1   (*a*) The John Snow pub in London where epidemiologists go to celebrate the heroics of John Snow's early epidemiological efforts to help stem a cholera epidemic. (*b*) A replica of the pump, with pump handle attached, a monument dedicated to Dr. Snow in July 1992. At the time of the epidemic, he was so convinced that the disease was being carried by water from the pump that he had the pump handle removed. Koch isolated and identified the cholera vibrio about 30 years later. Courtesy of Kathryn Foxhall.

(a)

(b)

EXERCISE

32

Epidemiology: A *Staphylococcus* Carrier Study

Getting Started

Nosocomial (hospital-acquired) infections and epidemics are common—they account for a considerable proportion of infections in hospital patients. This is not surprising, since these patients comprise a highly susceptible group of people. Moreover, the hospital is an environment in which procedures or treatment may have the effect of reducing normal body resistance to infection, and in which the use of antibiotics has fostered the development of drug resistant strains. What may come as a surprise to health science students is the degree to which hospital personnel themselves can be **carriers** and transmitters of pathogens that may also be resistant to antibiotics in many cases. In this exercise, you find out how many of the students in your class are carriers of the potential pathogen *Staphylococcus aureus*, and on what part of the body these bacteria occur.

Since penicillin and other antibiotics have been used increasingly over the last 50 years to treat infections, many hospital-isolated strains are resistant to one or more of the antibiotics to which the original *Staphylococcus* strains were formerly susceptible. Comparison of your antibiogram (antibiotic susceptibility pattern) to that of the reference strain demonstrates how these antibiograms have changed with the use of antibiotics.

When a number of patients in a given hospital have *Staphylococcus aureus* infections, it is often difficult to determine if a true epidemic exists, because so many people are routine carriers of these organisms on their body. However, by identifying different strains within the single species, it is possible to determine the existence of a true epidemic arising from a single source. If all isolated organisms come from the same strain, this is strong evidence they are from the same source; if they are of different strains, then the infections are probably not directly related.

Variations in susceptibility to antibiotics (the antibiogram), and in the production of **hemolysin** and **coagulase** can be studied to demonstrate different strains of *S. aureus*. In clinical studies, susceptibility of staphylococci to infection with different bacteriophages can be used for strain differentiation. The greater the number of characteristics studied, the more accurate strain identification becomes.

In this exercise, you prepare streak plates on mannitol salt agar (a **selective and differential medium** used for isolation of pathogenic staphylococci) of swabs taken from three areas of the body: the throat, nose, and skin.

Mannitol salt agar contains 7.5% NaCl which inhibits most organisms except the salt tolerant skin flora. Both *Staphylococcus* and *Micrococcus* can grow on this medium. The carbohydrate mannitol differentiates between mannitol fermenters and nonfermenters. *Micrococcus* and *Staphylococcus epidermidis* cannot ferment mannitol, but *Staphylococcus aureus* can. Mannitol fermentation produces acid products dropping the pH and turning the phenol red indicator yellow. Therefore yellow colonies on mannitol salt that have turned the agar yellow are possibly *Staphylococcus aureus*. *Micrococcus* frequently have yellow colonies, but they do not change the color of the agar (see color plate 27).

If a mannitol-fermenting colony is isolated from any source, you test an isolate for its ability to produce hemolysin and coagulase, and for resistance to six antibiotics to which the typical reference strain of *S. aureus* is susceptible.

The results of these studies are reported in table 32.1 of the Laboratory Report. From this table an Information Sheet is prepared and turned in to the instructor for tabulation with those of the other students. A summary of the tabulated data appears in table 32.2. With the latter data, you will be able to answer the epidemiological questions suggested in the Questions section of the Laboratory Report.

Definitions

Carrier. A person who harbors infections and inconspicuously spreads them to others.

Coagulase. An enzyme secreted by *S. aureus* that clots plasma. It contributes to virulence and to forming a fibrin wall that surrounds staph lesions.

Differential medium. A growth medium designed to distinguish one kind of organism from another based on appearance of the colony.

Hemolysin. Any biological agent capable of lysing red blood cells with concomitant release of hemoglobin. Examples include certain exotoxins and complement related reactions.

Selective medium. A growth medium designed to favor the growth of certain microbes and inhibit the growth of undesirable competitors.

Objectives

1. To enable you to participate actively in an epidemiological investigation, wherein you can use your microbiological skills to obtain new information (see Laboratory Report Questions 1 through 4).
2. To test three hypotheses about staphylococcal carriers (see Laboratory Report Questions 5 through 7).

References

Mausner, J. S., Kramer, Shira M. et al. *Epidemiology, an introductory text.* 2nd ed. Philadelphia: W. B. Saunders Co., 1985.

Nester et al. *Microbiology: A human perspective*, 4th ed., 2004. Chapter 4, Section 4.5 and Chapter 20, Section 20.1.

Materials

Per team of two students

TS broth 24-hour 37°C reference culture of *S. aureus* with its antibiogram checked previously by instructor.

Mannitol salt agar plates with phenol red indicator, 7

Sterile swabs, each in a sterile test tube, 9

Tongue depressors, 2

Tubes of sterile water, 2

Blood agar plates, 2

Tubes of TS broth, 4

Tubes containing 0.5 ml of coagulase plasma, 2

Sterile Pasteur pipets, 3

Mueller-Hinton agar, 3 plates

A dropper bottle containing *fresh* 3% hydrogen peroxide

Forceps, 2

mm ruler

Antibiotic discs, 3 of each of the following: penicillin, 10 μg erythromycin, 15 μg streptomycin, 10 μg tetracycline, 30 μg sulfanilamide, 300 μg chloramphenicol, 30 μg

Procedure

Safety Precautions: *S. aureus* can cause wound infections, food poisoning, and toxic shock syndrome.

First Session

1. Each student assembles three mannitol salt agar plates; label one plate "nose," another "throat," and the last plate "skin." Write your name on each plate.
2. Each student labels three test tubes containing swabs "nose," "throat," and "skin," respectively.
3. Using aseptic technique, have your partner moisten the swab from your tube labeled "nose" in a tube of sterile water, sponge it almost dry against the inside of the tube, and insert it into your anterior nostril by gentle rotation, and return the swab to the empty tube.
4. Repeat step 3 with the swab labeled "throat"; with a tongue depressor, depress the tongue, then rub the swab firmly over the back of the throat and tonsillar region (see figure 23.2). Return the swab to the empty tube.

5. Repeat step 3 with the swab labeled "skin"; rub it over the skin surface located between your fingers and finger tips; return the swab to the empty tube.
6. Streak each swab over ⅓ of the surface of the appropriately labeled mannitol salt agar plate. Discard used swabs in a hazardous waste container. Continue to streak with a loop to obtain isolated colonies.
7. Repeat steps 3 through 6 for your partner.
8. Streak a loopful of the *S. aureus* reference culture on the surface of a mannitol salt agar plate (only one reference culture per team is necessary).
9. Invert plates and incubate at 37°C until next session.

Second Session

1. Examine the reference strain mannitol salt agar plate and note the appearance of typical *S. aureus* colonies (usually large, opaque colonies); any with mannitol fermentation show yellow halos (see color plate 27). Look for similar colonies on your nose, throat, and skin mannitol salt agar plates. Record their mannitol fermentation results (+ or −) in part 1 of the Laboratory Report.
2. Select several typical appearing staph colonies from one or more of your body isolate plates and test for catalase production (see figure 23.3 for test method). Include a positive control from your reference mannitol salt agar plate.

   Note: A positive catalase test is only suggestive of the presence of *S. aureus*, since all staphylococci exhibit a positive catalase test. If all the colonies from your body culture plates are catalase negative, state so in the Laboratory Report, and fill out and turn in the Information Sheet to your instructor. This occurrence completes your laboratory portion of the study.
3. Twenty-four hours before the next laboratory session, subculture a colony from your body isolate mannitol salt agar plate that is catalase and mannitol positive. Also subculture the *S. aureus* reference strain. They should be subcultured in two tubes of TS broth and on a plate of blood agar. Both your body culture

and the reference strain can be streaked on one blood agar plate by dividing the plate into halves. Label the tubes and plates, and incubate at 37°C for 24 hours.
4. Determine the coagulase activity of another catalase and mannitol positive body plate colony, and the reference colony culture. Use the method outlined in exercise 22. Record your results in table 32.1.

Note: If you have only one catalase and mannitol positive colony on your chosen body plate, delay the coagulase test until the next session and then use a colony from the blood agar plate for the coagulase test.

Third Session

1. Record the presence or absence of hemolysis on the blood agar plates in table 32.1. Compare it for similarity with the reference strain of *S. aureus*.
2. If the coagulase test is negative, report your culture as negative for *S. aureus*. Fill out the Information Sheet and turn it in to your instructor.
3. If the coagulase test is positive, set up an antibiotic susceptibility test. The six antibiotic discs to be used are listed in the Materials section. The inocula to be used are the three TS broth cultures prepared in the Second Session. The procedure is as follows:
   a. With a permanent marking pen, divide the bottom of a petri dish containing Mueller-Hinton agar into six pie-shaped sections (see figure 14.3a). Repeat the marking procedure with the remaining two dishes of Mueller-Hinton agar. Label one dish Reference culture. The remaining two dishes represent you and your partner's body cultures.
   b. Inoculate the reference plate by moistening a sterile swab with the TS broth reference culture, and spreading it uniformly over the plate surface by moving the swab back and forth in three directions (see figure 14.3b). Repeat the inoculation procedure with the remaining two plates using your two TS broth cultures as inocula.

Note: In order to have some semblance of validity, the procedures used to prepare and observe the antibiotic susceptibility test should be similar to that used in exercise 14. Use table 14.2 to interpret your results.

c. Assemble the discs, and record the code of each in part 3 of the Laboratory Report to ensure correct interpretation of test results.

d. Heat-sterilize forceps by dipping in 95% alcohol and flaming. Air cool.

e. Remove one disc aseptically from container. Place gently in the center of one pie-shaped section of the reference plate culture (see figure 14.3*a*). Tap disc gently with forceps to fix it in position on the agar surface.

f. Continue placing the remaining five discs in the same way. Make certain that you sterilize the forceps after placing each disc, since there is a possibility of contaminating stock vials with resistant organisms or even occasionally with drug-dependent bacteria!

g. Repeat the procedure with your two body culture plates.

h. Invert and incubate the plates at 35°C for 48 hours.

Fourth Session

1. Using a mm ruler, measure the diameter of the zone of inhibition around each antibiotic disc, and record their diameters in table 32.1. Next consult table 14.2 and determine from the latter information if the cultures are susceptible (S) or resistant (R) to the antibiotic in question. Record S or R in the appropriate square of table 32.1.

   Note: The reference culture is expected to be susceptible to all six antibiotics.

2. Fill out the Information Sheet of the Laboratory Report. Tear out and return it to the instructor for tabulation of data to be inserted in table 32.2.

Name _____ Date _____ Section _____

EXERCISE

Laboratory Report:
Epidemiology: A *Staphylococcus* Carrier Study

Results

1. Record the initial results you obtained from your three body cultures on mannitol salt agar.

Note: At least one colony should show good colony growth with a yellow color change indicating a positive test for mannitol fermentation.

Record each culture as mannitol positive (+) or negative (–):

Nose_____   Throat_____   Skin_____

2. Catalase test results (+ or –):

Nose_____   Throat_____   Skin_____

3. Record the code of each antibiotic disc:

Penicillin_____   Erythromycin_____

Streptomycin_____   Tetracycline_____

Sulfanilamide_____   Chloramphenicol_____

4. Record results obtained with the one strain you chose to study in table 32.1. Also indicate deviation of any of the results obtained for your strain from the reference strain.

Table 32.1   Test Results Obtained with Reference *S. aureus* Culture and Mannitol Salt Agar Positive Body Culture

Isolate Source Tested	Mannitol Fermentation (+ or –)	Beta-hemolysis (+ or –)	Coagulase (+ or –)	Antibiotic Susceptibility (S or R)					
				Peni-cillin	Strepto-mycin	Tetra-cycline	Chloram-phenicol	Erythro-mycin	Sulfanil-amide
Your Strain									
Reference Strain									
Deviation from Reference Strain									

NOTES:

Name _____ Date _____ Section _____

Instructor Information Sheet for Tabulation of *Staphylococcus* Carrier Study

Which sources showed growth and fermentation on mannitol salt agar?

Throat____   Nose____   Skin____

Do you work in a clinical setting this quarter?

Yes____ No____

If so, where?_____

Have you taken antibiotics this quarter?

Yes____ No____

If so, which?_____

Isolate Source Tested	Mannitol Fermentation (+ or −)	Beta-hemolysis (+ or −)	Coagulase (+ or −)	Antibiotic Susceptibility (S or R)					
				Peni-cillin	Strepto-mycin	Tetra-cycline	Chloram-phenicol	Erythro-mycin	Sulfanil-amide
Your Strain									
Reference Strain									
Deviation from Reference Strain									

NOTES:

5. Based on the information for the class given to you by your instructor, complete table 32.2.

Table 32.2   Classroom Summary of *S. aureus* Epidemiological Study

	Health Professionals (students working in clinical setting)		General Population (students not working in clinical setting)		Total Class Results	
	Number	% of sample	Number	% of sample	Number	% of sample
S. aureus carriers (positive cultures)						
Noncarriers (negative cultures)						
Total						
Carriers of susceptible strains						
Carriers of resistant strains						
Total						
Carriers of strainlike reference strain						
Carriers of deviant strains						
Total						
Different *types* of deviant strains						

Total number and % of mannitol positive strains from:

Throat _____ Nose _____ Skin _____ Total _____

(**Note:** This number might be larger than the class, because any student might have any number of sources positive.)

Questions

1. What proportion of students are carriers of the potential pathogen *Staphylococcus aureus?*

2. What parts of the body harbor this pathogen, and which parts carry it most commonly?

3. To what extent has *S. aureus* acquired resistance to antibiotics to which it was originally susceptible?

4. How many different strains can be isolated from the student population that deviate from the "typical" *S. aureus?*

5. Do working health professionals have a higher carrier rate than the general population?

6. Do the strains of S. *aureus* carried by health professionals have a higher proportion of resistant strains than those isolated from the general population?

7. Is there a greater number of different strains of S. *aureus* among health professional carriers than among the general population?

NOTES:

EXERCISE

33

Bacteriological Examination of Water: Multiple-Tube Fermentation and Membrane Filter Techniques

Getting Started

Water, water, everywhere Nor any drop to drink.
        The Ancient Mariner Coleridge, 1796

This rhyme refers to sea water, undrinkable because of its high salt content. Today the same can be said of freshwater supplies, polluted primarily by humans and their activities. A typhoid epidemic, dead fish on the beach, and the occurrence of a red tide are all visible evidence of pollution. Primary causes of pollution include dumping of untreated (raw) sewage and inorganic and organic industrial wastes, and fecal pollution by humans and animals of both fresh and groundwater. In the United States, sewage and chemical wastes are in regression largely as a result of the passage of federal and local legislation requiring a minimum of secondary treatment for sewage and the infliction of severe penalties for careless dumping of chemical wastes.

Fecal pollution by humans and animals is more difficult to control particularly as the supply of water throughout the world becomes more critical. In some parts of the world, particularly in Third World countries, it is estimated that over 12,000 children die every day from diseases caused by waterborne fecal pollutants. Examples of such diseases are cholera, typhoid fever, bacterial and amoebic dysentery, and viral diseases such as polio and infectious hepatitis. Most of the inhabitants of these countries are in intimate contact with polluted water because they not only drink it, but also bathe, swim, and wash their clothes in it.

The increased organic matter in such water also serves as a substrate for **anaerobic** bacteria, thereby increasing their numbers in relation to the **aerobic** bacteria originally present (figure 33.1). Nuisance bacteria, such as *Sphaerotilus natans*, a large, rod-shaped organism that grows in chains and forms an external sheath (figure 33.2) are able to adhere to

Figure 33.1 Diversity of microbes found in pure to very polluted water. Note the change from aerobic to anaerobic microbes as the water becomes more polluted. Courtesy of Settlemire and Hughes. *Microbiology for Health Students*, Reston Publishing Co., Reston, Virginia.

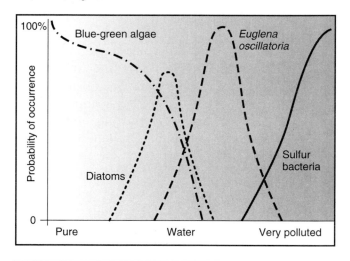

Figure 33.2 *Sphaerotilus* species, a sheathed bacterium that often produces masses of brownish scum beneath the surface of polluted streams. Phase-contrast photomicrograph courtesy of J. T. Staley and J. P. Dalmasso.

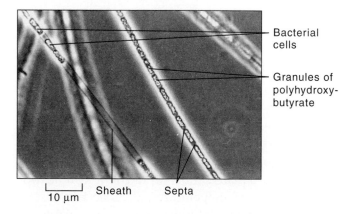

the walls of water pipes. This can eventually cause fouling of the pipes, thereby reducing the carrying capacity of the pipes.

Fortunately, microbes are also beneficial in water purification. In smaller sewage treatment plants, raw sewage is passed through a slow sand filter, wherein

Figure 33.3 Human fecal specimen illustrations showing trophozoite (*a*) and cyst (*b*) forms of *Giardia lamblia*, a waterborne protozoan pathogen that inhabits the intestinal tract of certain warm-blooded animals. The cyst form is resistant to adverse environments and is the form released with fecal material. (*a*) Pear-shaped trophozoite. Electron microscopy, iodine-stained, magnification × 2,900, 1 micron = 2.9 mm. Note flagellate appendages for locomotion. (*b*) Smooth ovoid cysts. Scanning electron microscopy, magnification × 2,900, 1 micron = 2.9 mm. Cysts are embedded in a mat of debris, bacteria, and fecal material. (*a*) © J. Paulin/Visuals Unlimited (*b*) From D. W. Luchtel, W. P. Lawrence, and F. B. De Walle, "Electron Microscopy of Giardia lamblia Cysts," *Applied and environmental microbiology*, 40:821–832, 1980. © American Society for Microbiology.

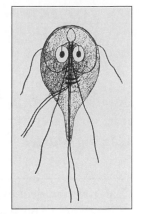

Drawing of trophozoite

microorganisms present in the sand are able to degrade (metabolize) organic waste compounds before the effluent is discharged. Sewage effluent is finally chlorinated to further reduce fecal microbial contaminants. The development of sewage treatment plants and the control of raw sewage discharge reduced the annual typhoid fever death rate in the United States from about 70 deaths per 100,000 population to nearly zero. However, the potential danger of pollution is always present. In 1973, an epidemic outbreak of typhoid fever occurred in Florida, and in 1975 residents of the city of Camas, Washington were inundated by intestinal disorders that were traced to fecal pollution of the water supply by beavers infected with the protozoan *Giardia lamblia* (figure 33.3).

Two microbiological methods commonly used for determining whether a given sample of water is polluted are:

1. The determination of the total number of microorganisms present in the water. The plate count method provides an indication of the amount of organic matter present. In view of the great diversity in microbial physiology, no single growth medium and no single set of cultural conditions can be found that satisfy universal microbial growth. Hence, a choice had to be made. Experience taught that standard plate counts on nutrient agar at two incubation temperatures, 20°C and 35°C, provided a useful indication of the organic pollution load in water.

2. The determination of fecal contamination, and hence the possible presence of pathogens, with the help of suitable **indicator organisms.** Indicator organisms are normally nonpathogenic, always occur in large

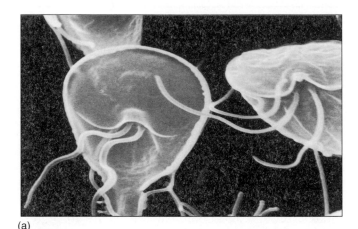

(a)

(b)

quantities in feces, and are relatively easy to detect as compared to detection of waterborne pathogens. The diagnosis of the latter is usually more complicated and time-consuming, and thus less suited for routine investigations. Assuming that in cold surface waters the pathogens are dying off faster than the indicator organisms employed, the absence of the latter or their presence in very low numbers guaranteed, in most cases, the absence of pathogens. Recently, however, better techniques and procedures have been developed for the detection of pathogenic bacteria and viruses. In certain cases, these techniques have indicated the presence of pathogens in the absence of indicator organisms hitherto relied upon, showing increased resistance of pathogens to the aqueous environment. Similar findings have been found with respect to resistance to chlorination. Such results suggest that some of the principles and methods in conventional water examination are of questionable value.

Nevertheless, conventional methods for detection of fecal contamination are still widely employed. Some will be described here. In general, these procedures are the same ones described by the American Public Health Association in *Standard Methods for the Examination of Water and Wastewater.*

The indicator organisms most widely used belong to the so-called **coliform group.** This includes all aerobic and facultative anaerobic, Gram-negative, non-spore-forming, rod-shaped bacteria that ferment lactose with gas formation within 48 hours at 37°C, and comprises *Escherichia coli* (10^6 to 10^9 cells/g of feces) together with a number of closely related organisms (see exercise 24). Noncoliforms that are sometimes employed, primarily for confirmation, include *Streptococcus faecalis*, some related species, and in Great Britain, *Clostridium perfringens*, which is also called C. *welchii.*

The presence of *E. coli* in water from sources such as reservoirs suggests that chlorination is inadequate. Current standards for drinking water state that it should be free of coliforms and contain no more than 10 other microorganisms per ml.

Two of the most important methods applied to detect coliform organisms are the multiple-tube fermentation technique and the membrane filter technique.

Multiple-Tube Fermentation Technique

This technique employs three consecutive tests: first a presumptive test; if the first test is positive, then a confirmed test; and finally a completed test (figure 33.4 provides a pictorial description of these tests).

Presumptive Test This test, a specific enrichment procedure for **coliform bacteria,** is conducted in fermentation tubes filled with a **selective growth medium** (MacConkey lactose broth), which contain inverted Durham tubes for detection of fermentation gas (see figure 33.4).

The main selective factors found in the medium are lactose, sometimes a **surfactant** such as Na-lauryl sulfate or Na-taurocholate (bile salt), and often a pH indicator dye for facilitating detection of acid production, such as bromcresol purple or brilliant green. The selective action of lactose occurs because many bacteria cannot ferment this sugar, whereas coliform bacteria and several other bacterial types can ferment it. The surfactant and dye do not inhibit coliform bacteria, whereas many other bacteria, such as the spore formers, are inhibited. The original surfactant used by MacConkey was bile salt because *E. coli* is adapted to growing in the colon, whereas nonintestinal bacteria generally are not.

The formation of 10% gas or more in the Durham tube within 24 to 48 hours, together with turbidity in the growth medium constitutes a positive presumptive test for coliform bacteria, and hence for the possibility of fecal pollution. The test is presumptive only, because under these conditions several other types of bacteria can produce similar results.

The presumptive test also enables quantitation of the bacteria present in the water sample. The test, described as the most probable number test (MPN), is also useful for counting bacteria that reluctantly form colonies on agar plates or membrane filters, but grow readily in liquid media. In principle, the water sample is diluted so that some of the broth tubes contain a single bacterial cell. After incubation, some broth tubes show growth with gas, whereas others do not. The total viable count is then determined by counting the portion of positive tubes and referring this data to a statistical MPN table used for calculating the total viable bacterial count (see Procedure, table 33.1).

Figure 33.4 Standard methods procedure for the examination of water and wastewater and for use in determining most probable number (MPN).

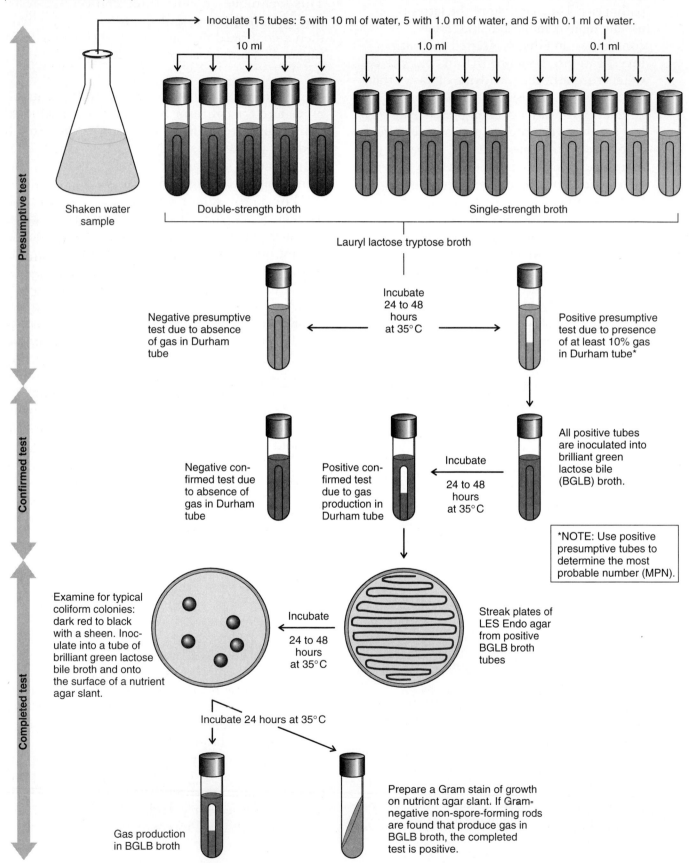

Inoculate 15 tubes: 5 with 10 ml of water, 5 with 1.0 ml of water, and 5 with 0.1 ml of water.

10 ml

1.0 ml

0.1 ml

Presumptive test

Shaken water sample

Double-strength broth

Single-strength broth

Lauryl lactose tryptose broth

Negative presumptive test due to absence of gas in Durham tube

Incubate 24 to 48 hours at 35°C

Positive presumptive test due to presence of at least 10% gas in Durham tube*

Confirmed test

Negative confirmed test due to absence of gas in Durham tube

Positive confirmed test due to gas production in Durham tube

Incubate 24 to 48 hours at 35°C

All positive tubes are inoculated into brilliant green lactose bile (BGLB) broth.

*NOTE: Use positive presumptive tubes to determine the most probable number (MPN).

Completed test

Examine for typical coliform colonies: dark red to black with a sheen. Inoculate into a tube of brilliant green lactose bile broth and onto the surface of a nutrient agar slant.

Incubate 24 to 48 hours at 35°C

Streak plates of LES Endo agar from positive BGLB broth tubes

Incubate 24 hours at 35°C

Gas production in BGLB broth

Prepare a Gram stain of growth on nutrient agar slant. If Gram-negative non-spore-forming rods are found that produce gas in BGLB broth, the completed test is positive.

Confirmed Test This test serves to confirm the presence of coliform bacteria when either a positive or doubtful presumptive test is obtained. A loopful of growth from such a presumptive tube is transferred into a tube of brilliant green bile 2% broth and incubated at 35°C for 48 hours. This is a selective medium for detecting coliform bacteria in water, dairy, and other food products. In order to do so, the correct concentration of the dye (brilliant green) and bile must be present. If it is too concentrated, coliform growth can also be inhibited. Bile is naturally found in the intestine where it serves a similar purpose, encouraging growth of coliform bacteria while discouraging growth of other bacteria. A final selective agent in the medium is lactose. The broth tube also contains a Durham tube to detect gas production. The presence of gas in the Durham tube after incubating for 24 to 48 hours constitutes a positive confirmed test.

Completed Test This test helps to further confirm doubtful and, if desired, positive confirmed test results. The test is in two parts:

1. A plate of LES Endo agar (or Levine's EMB agar) is streaked with a loopful of growth from a positive confirmed tube, and incubated at 35°C for 18–24 hours. Typical coliform bacteria (*E. coli* and *Enterobacter aerogenes*) exhibit good growth on this medium and form red to black colonies with a sheen. *Salmonella typhi* exhibits good growth but the colonies are colorless. *S. aureus* growth is inhibited altogether.

2. Next a typical coliform colony from an LES Endo agar plate is inoculated into a tube of brilliant green bile broth and on the surface of a nutrient agar slant. They are then incubated at 35°C for 24 hours. A Gram stain is then prepared from growth present on the nutrient agar slant. The presence of gas in the brilliant green bile broth tube and Gram-negative non-spore-forming rods constitutes a positive completed test for the presence of coliform bacteria, which, in turn, infers possible contamination of the water sample with fecal matter.

Membrane Filter Technique

For this technique, a known volume of water sample (100 ml) or of its dilutions is filtered by suction through a sterile polycarbonate or nitrocellulose acetate membrane filter. The filter is very thin (150 μm), and has a pore diameter of 0.45 μm. The precision manufacture of the filter is such that bacteria larger than 0.47 μm cannot pass through it. Filters with printed grid patterns are available for facilitating colony counting.

Once the water is filtered, the filter disc is aseptically transferred to the surface of a wetted pad contained in a petri dish. The pad is soaked with Endo broth MF on which coliform colonies will appear to be pink to dark red in color and possess a golden green metallic sheen. After incubation at 35°C for 24 hours, the filter disc is examined for characteristic coliform colonies and their number. From this number, one can calculate the total number of coliform bacteria present in the original water sample. For additional confirmation, the procedure for the completed multiple-tube fermentation test can be applied.

The membrane filter method yields accurate results if the coliform colony count is in the range of 30 to 300 organisms per filter disc. Unknown samples require that various dilutions be examined. Turbidity of the sample constitutes a serious obstacle in cases where dilutions, yielding coliform counts within the desired range, contain sufficient suspended matter to clog the filter before the required volume of water has passed through.

The advantages of the membrane filter technique over the multiple-tube fermentation test are: (1) better reproducibility of results; (2) greater sensitivity, because larger amounts of water can be used; and (3) shorter time (one-quarter the time) for obtaining results. This method has been recognized by the United States Public Health Service for detection of coliforms in water.

Definitions

Aerobic bacteria. Microbes that grow and multiply in the presence of free gaseous oxygen.

Anaerobic bacteria. Microbes that grow best, or exclusively in the absence of free oxygen.

Coliform bacteria. A collective term for bacteria that inhabit the colon, are Gram negative and ferment lactose (see page 287).

Indicator organism. A nonpathogenic organism whose presence when detected in water or sewage serves as an indicator of possible pollution with pathogens.

Selective growth medium. A growth medium that contains substances that inhibit the growth of certain organisms but not others.

Surfactant. A surface-active agent that forms a water-soluble common boundary between two substances. Examples include detergents and wetting agents.

Objectives

1. To introduce you to the use of a multiple-tube fermentation technique for detecting the presence and number of coliform pollution indicator organisms present in water samples.
2. To introduce you to the use of an alternate method, the membrane filter technique, for detecting the presence and number of coliform bacteria in water samples.

References

McKinney, R. E. *Microbiology for sanitary engineers,* New York: McGraw-Hill Book Co., 1972.

Nester et al. *Microbiology: A human perspective,* 4th ed., 2004. Chapter 31, Section 31.2.

Standard methods for the examination of water and wastewater, 18th ed. The American Water Works Association, 6666 West Quincy Ave., Denver, Co 80235.

Materials

Multiple-Tube Fermentation Technique (per student), see figure 33.4

Test tubes, 50 ml, containing 10 ml of double strength lauryl sulfate (lauryl lactose tryptose) broth plus Durham tubes, 5

Small test tubes containing 10 ml of single strength lauryl sulfate broth plus Durham tubes, 10

Sterile 10-ml pipet, 1

Sterile 1-ml pipet calibrated in 0.1 ml units, 1

Brilliant green bile 2% broth plus Durham tubes, 2 tubes

LES Endo agar plate, 1

Nutrient agar slant, 1

Sterile 100-ml screw cap bottle for collecting water sample, 1

Membrane Filter Technique (demonstration), see figure 33.5

A 1-liter side-arm Erlenmeyer flask, 1

Sterile membrane filter holder assembly, two parts wrapped separately (see figure 33.5, frames 2 and 3), 1 unit

A metal clamp for clamping filter funnel to filter base

Sterile membrane filters, 47 mm diameter, 0.45 μm pore size

Forceps, 1 pair

Sterile 50-mm diameter petri dishes, 3

Absorbent filter pads, 3

Tube containing 10 ml of sterile Endo MF broth, 1

Sterile 90-ml water blanks, 2

Erlenmeyer flasks containing 25 ml of sterile water, 6

Sterile 10-ml pipets, 2

Vacuum pump or Venturi vacuum system

Water sample for coliform analysis, 100 ml

Procedure

Multiple-Tube Fermentation Technique

Note: If desired, your instructor may ask you to bring a 50- to 100-ml sample of water from home, a nearby stream, a lake, or some other location for analysis. When taking a tap sample, the orifice of the water tap should be flamed before being opened. After opening, allow water to run for 5 to 10 minutes with the tap in the same position to prevent loosening of bacteria from inside the tap. Next, using aseptic technique, open a sterile bottle (obtained from the instructor beforehand) and collect a sample. If the sample cannot be examined within 1–2 hours, keep refrigerated until ready for use.

First Session (Presumptive Test)

1. Shake water sample. Aseptically pipette 10-ml portions of the sample into each of the five large tubes containing 10-ml aliquots of double-strength lauryl sulfate broth. Next, with a 1-ml pipet, transfer 1-ml portions of the water sample into five of the smaller tubes, and 0.1-ml portions into the remaining five small tubes of lauryl sulfate broth. Be sure to label the tubes.
2. Incubate the test tubes for 48 hours at 37°C.

Second and Third Sessions (Presumptive and Confirmed Tests)

1. Observe the tubes after 24 hours for gas production by gently shaking the tubes. If after shaking gas is not evident in the Durham tube, reincubate the tube for an additional 24 hours. Record any positive results for gas production in table 33.2 of the Laboratory Report.
2. Observe the tubes for gas production and turbidity after 48 hours of incubation. If neither gas nor turbidity are present in any of the tubes, the test is negative. If turbidity is present but no gas, the test may be doubtful since growth without gas may mean inhibition of coliform bacteria by noncoliform bacteria with shorter generation times. If at least 10% of the tube is filled with gas, the test is positive for coliform bacteria. Record your results in table 33.2 of the Laboratory Report.
3. MPN determination. Using your fermentation gas results in table 33.2, determine the number of tubes from each set containing 10% gas or more. Determine the MPN by consulting table 33.1; for example, if you had gas in two of the first five tubes, in two of the second five tubes, and none in the third three tubes, your test readout would be 2-2-0. Table 33.1 shows that the MPN for this readout would be 9. Thus, your water sample would contain nine organisms per 100-ml water with a 95% statistical probability of there being between three and twenty-five organisms.

Note: If your readout for the series is 0-0-0, it means that the MPN is less than two organisms per 100 ml of water. Also, if the readout is 5-5-5 it means the MPN is greater than 1,600 organisms/100 ml water. In the latter instance, what procedural modification would be required to obtain a more significant result? Report your answer in question 10 of the Questions section in the Laboratory Report.

4. The confirmed test should be administered to all tubes demonstrating either a positive or doubtful presumptive test. Inoculate a loopful of growth from each tube showing gas or dense turbidity into a tube of brilliant green lactose bile 2% broth. Incubate the tube(s) at 37°C for 24 to 48 hours.

Note: For expediency, your instructor may wish you to inoculate only one tube. If so, for the inoculum use the tube of lauryl sulfate broth testing positive with the least inoculum of water.

Fourth Session (Confirmed and Completed Tests)

1. Examine the brilliant green lactose bile 2% tube(s) for gas production. Record your findings in the confirmed test section of the Laboratory Report.
2. Streak a loopful of growth from a positive tube of brilliant green lactose bile 2% broth on the surface of a plate containing LES Endo agar. Incubate at 37°C for 24 hours.

Fifth Session (Completed Test)

1. Examine the LES Endo agar plate(s) for the presence of typical coliform colonies (dark red to black with a sheen). Record your findings in the completed test section of the Laboratory Report.
2. With a loop, streak a nutrient agar slant with growth obtained from a typical coliform colony found on the LES Endo agar plate. Also inoculate a tube of brilliant green lactose bile 2% broth with growth from the same colony. Incubate the tubes at 37°C for 24 hours.

Sixth Session (Completed Test)

1. Examine the brilliant green lactose bile 2% broth tube for gas production. Record your result in the completed test section of the Laboratory Report.

Table 33.1 MPN Index and 95% Confidence Limits for Various Combinations of Positive Results When Five Tubes Are Used Per Dilution (10 mL, 1.0 mL, 0.1 mL)

Combination of Positives	MPN Index/ 100 mL	95% Confidence Limits Lower	95% Confidence Limits Upper	Combination of Positives	MPN Index/ 100 mL	95% Confidence Limits Lower	95% Confidence Limits Upper
0-0-0	<2	—	—	4-3-0	27	12	67
0-0-1	3	1.0	10	4-3-1	33	15	77
0-1-0	3	1.0	10	4-4-0	34	16	80
0-2-0	4	1.0	13	5-0-0	23	9.0	86
1-0-0	2	1.0	11	5-0-1	30	10	110
1-0-1	4	1.0	15	5-0-2	40	20	140
1-1-0	4	1.0	15	5-1-0	30	10	120
1-1-1	6	2.0	18	5-1-1	50	10	150
1-2-0	6	2.0	18	5-1-2	60	30	180
2-0-0	4	1.0	17	5-2-0	50	20	170
2-0-1	7	2.0	20	5-2-1	70	30	210
2-1-0	7	2.0	21	5-2-2	90	40	250
2-1-1	9	3.0	24	5-3-0	80	30	250
2-2-0	9	3.0	25	5-3-1	110	40	300
2-3-0	12	5.0	29	5-3-2	140	60	360
3-0-0	8	3.0	24	5-3-3	170	80	410
3-0-1	11	4.0	29	5-4-0	130	50	390
3-1-0	11	4.0	29	5-4-1	170	70	480
3-1-1	14	6.0	35	5-4-2	220	100	580
3-2-0	14	6.0	35	5-4-3	280	120	690
3-2-3	17	7.0	40	5-4-4	350	160	820
4-0-0	13	5.0	38	5-5-0	240	100	940
4-0-1	17	7.0	45	5-5-1	300	100	1300
4-1-0	17	7.0	46	5-5-2	500	200	2000
4-1-1	21	9.0	55	5-5-3	900	300	2900
4-1-2	26	12	63	5-5-4	1600	600	5300
4-2-0	22	9.0	56	5-5-5	≥1600	—	—
4-2-1	26	12	65				

From *Standard Methods for the Examination of Water and Wastewater*, 18th edition. Copyright 1992 by the American Public Health Association, the American Water Works Association, and the Water Environment Federation. Reprinted with permission.

2. Prepare a Gram stain of some of the growth present on the nutrient agar slant. Examine the slide for the presence of Gram-negative, non-spore-forming rods. Record your results in the completed test section of the Laboratory Report. The presence of gas and of Gram-negative, non-spore-forming rods constitutes a positive completed coliform test.

Membrane Filter Technique

First Session

1. Shake the water sample. Prepare two dilutions by transferring successive 10-ml aliquots into 90-ml blanks of sterile water (10^{-1} and 10^{-2} dilutions).

Note: Reshake prepared dilutions before using.

2. Assemble the filter holder apparatus as follows (figure 33.5):
   a. Using aseptic technique, unwrap the lower portion of the filter base, and insert a rubber stopper.
   b. Insert the base in the neck of the 1-liter side-arm Erlenmeyer flask (figure 33.5, frame 2).
   c. With sterile forceps (sterilize by dipping in alcohol and flaming them in the flame of the Bunsen burner), transfer a sterile membrane filter onto the sintered glass or plastic surface of the filter holder base (figure 33.5, frames 1 and 2). Make certain the membrane filter is placed with the ruled side up.
   d. Aseptically remove the covered filter funnel from the butcher paper, and place the lower surface on top of the membrane filter. Clamp the filter funnel to the filter base with the clamp provided with the filter holder assembly (figure 33.5, frame 3).
3. Prepare 3 plates of Endo medium by adding 2-ml aliquots of the tubed broth to sterile absorbent pads previously placed aseptically with sterile tweezers on the bottom of the three petri dishes (figure 33.5, frames 4 and 5).
4. Remove the aluminum filter cover and pour the highest water dilution (10^{-2}) into the funnel (figure 33.5, frame 6). Assist the filtration process by turning on the vacuum pump or Venturi system.
5. Rinse the funnel walls with two 25-ml aliquots of sterile water.
6. Turn off (break) the vacuum and remove the filter holder funnel. Using aseptic technique and sterile tweezers, transfer the filter immediately to the previously prepared petri dish (figure 33.5, frames 7 and 8). Using a slight rolling motion, center the filter, grid side up, on the medium-soaked absorbent pad. Take care not to trap air under the filter as this will prevent nutrient media from reaching all of the membrane surface (figure 33.5, frame 9).
7. Reassemble the filter apparatus with a new membrane filter, and repeat the filtration process first with the 10^{-1} water sample, and finally with a 100-ml aliquot of the undiluted water sample.
8. Label and invert the petri dishes to prevent any condensate from falling on the filter surface during incubation. Incubate plates for 24 hours at 37°C.

Second Session

1. Count the number of coliform type bacteria by using either the low power of the microscope or a dissecting microscope. Count only those colonies that exhibit a pink to dark red center with or without a distinct golden green metallic sheen.
2. Record the number of colonies found in each of the three dilutions in table 33.3 of the Laboratory Report.

Figure 33.5 Analysis of water for fecal contamination. Cellulose acetate membrane filter method. (*1*) Sterile membrane filter (0.45 μm) with grid for counting is handled with sterile forceps. (*2*) The sterile membrane filter is placed on filter holder base with grid side up. (*3*) The apparatus is assembled. (*4*) Sterile absorbent pads are aseptically placed in the bottom of three sterile petri dishes. (*5*) Each absorbent pad is saturated with 2.0 ml of Endo broth. (*6*) A portion of well-mixed water sample is poured into assembled funnel and filtered by vacuum. (*7*) Membrane filter is carefully removed with sterile forceps after disassembling the funnel. (*8*) Membrane filter is centered on the surface of the Endo-soaked absorbent pad (grid side up) by use of a slight rolling motion. (*9*) After incubation, the number of colonies on the filter is counted. The number of colonies on the filter reflects the number of coliform bacteria present in the original sample.

EXERCISE

Laboratory Report: Bacteriological Examination
of Water: Multiple-Tube Fermentation
and Membrane Filter Techniques

Results

1. Multiple-Tube Fermentation Technique
   a. Record the results of the presumptive test in table 33.2.

Table 33.2   Presumptive Test for the Presence or Absence of Gas and Turbidity in Multiple-Tube Fermentation Media

| Water Sample Size (ml) | Presence of Gas and Turbidity* | | | | |
	Tube #1	Tube #2	Tube #3	Tube #4	Tube #5
10					
1					
0.1					

*Use a (+) sign to indicate gas and a circle (O) around the plus sign to indicate turbidity.

   b. Determine the MPN below:

   Test readout:_____   MPN:_____   95% Confidence Limits:_____

   c. Confirmed test results (gas production in brilliant green lactose bile 2% broth):

   | *Sample Number* | *Gas (+ or −)* | |
   | | 24 hours | 48 hours |

   d. Appearance of colonies on LES Endo agar:

   e. Completed test results:

   | *Sample Number* | *Gas (+ or −)* | *Gram Stain Reaction* |

2. Membrane filter technique

Table 33.3   Number of Coliform Colonies Present in Various Dilutions of the Water Sample

Undiluted Sample	10⁻¹ Dilution	10⁻² Dilution

a. Calculate the number of coliform colonies/ml present in the original water sample (show your calculations):

Questions

1. What nutritional means might be used to speed up the growth of the coliform organisms using the membrane filter technique?

2. Describe two other applications of the membrane filter technique.

3. Why not test for pathogens such as *Salmonella* directly rather than use an indicator organism such as the coliform bacteria?

4. Why does a positive presumptive test not necessarily indicate that the water is unsafe for drinking?

5. List three organisms that are apt to give a positive presumptive test.

6. Describe the purpose of lactose and Endo agar in these tests.

7. What are some limitations of the membrane filter technique?

8. Define the term *coliform*.

9. Briefly explain what is meant by presumptive, confirmed, and completed tests in water analysis.

10. See Note 3, MPN determination, on page 291 for the question.

*I*ntroduction to Biotechnology

Watson and Crick first proposed the structure of DNA in the early 1950s. In less than 50 years it became possible to isolate DNA, transfer specific genes to another organism, and determine the sequence of bases in the DNA of specific genes as well as entire bacterial genomes. These sequences can then be used to identify or classify an organism, and determine evolutionary relationships.

The techniques for manipulating DNA have many applications. A particular gene may be cut from the DNA of one organism using restriction enzymes and inserted into another organism, so that the action of the gene can be studied independently of the organism. Also it is a method for obtaining a product of a gene, usually a protein, in large quantities. For instance, the human gene for insulin can be cloned into yeast or bacterial DNA. These microorganisms can be grown in huge quantities and the insulin can be purified for people requiring it for the treatment of diabetes.

In these next two exercises you will apply some of the techniques used in biotechnology to identify DNA. First you will use restriction enzymes to produce patterns of DNA unique to each organism. In the second exercise you will use a computer database to identify an organism from its DNA sequences of the 16S rRNA gene.

NOTES:

EXERCISE 34

Identifying DNA with Restriction Enzymes

Getting Started

DNA from one organism can be distinguished from the DNA from another organism by the use of a type of enzyme called restriction enzymes. Restriction enzymes not only cut DNA at very specific sequences, but there are many different enzymes, each with their own different sequence. For example, the enzyme Hha I cuts at the sequence GCGC. In the two sequences of DNA below, each strand would be cut in two pieces because each strand contains GCGC, but the pieces would differ in size between the two sequences.

```
ATGGCTCAA GCGC TCACGGTAACTGCTGCATC-CCGTTATACGAGCTACT
TACCGAGTT CGCG AGT-GCCATTGACGACGTAGGGCAATATGCTCGATGA
```

```
ATATCGTTGAACTCCGTGTAGACT GCGC ACGTG-TTACAATCCACCAAGT
TATAGCAACTTGAGGCACATCTGA CGCG TGCACAATGTTAGGTGGTTCA
```

The location of the specific sequences varies from species to species and only identical strands of DNA will be cut into the same number and size of fragments. This is the basis for comparing DNA from different organisms and has many applications. For example in forensic (legal) investigations the DNA from blood stains can be compared with the DNA of a suspected murderer. In epidemiology investigations the DNA from a serious *E. coli* outbreak of diarrhea can be compared to the DNA isolated from other *E. coli* strains isolated from food to determine the source of infection.

Where do these enzymes come from? They are found in bacteria which use them to degrade foreign DNA that might enter their cell. Each cell methylates its DNA by adding a methyl group at a particular site and thereby prevents its own DNA from being degraded by its own restriction enzymes. Foreign DNA entering the cell does not have this specific pattern of methylation and the cell cleaves it, restricting its expression in the cell. Therefore these enzymes are called **restriction enzymes.** They can be isolated from bacteria and used in the laboratory for studying and manipulating DNA. The restriction enzymes are named after the bacteria's first letter of the genus and first two letters of the species. EcoRI is the first restriction enzyme from *Escherichia coli* strain **R**.

In the laboratory, samples of the DNA to be compared are mixed with a restriction enzyme and incubated until the enzymes have cleaved the DNA at the recognition site unique to the enzyme. Each sample of DNA contains pieces of DNA of specific lengths. How, then, do you determine the size of the fragments? This can be done by running the DNA on an **electrophoresis** agarose gel.

The gel is made from highly refined agar, called agarose (figure 34.1). Holes or wells are formed in the agarose when it is poured in the mold. The gel is covered with buffer and the DNA that has been cut is placed in the wells. A current is applied, and the negatively charged DNA moves to the positive pole. The smaller the pieces, the

Figure 34.1 Gel electrophoresis of three DNA samples including a size standard.

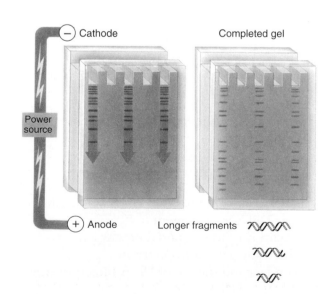

more quickly they move through the porous agarose gel. Since we cannot see the DNA in the gel, dyes are added to the sample that travel through the agar at about the same speed as the DNA, so we can have an idea how far the bands have traveled. If the gel runs too long, the samples will run off the bottom of the gel and diffuse into the **buffer.**

Ethidium bromide can be added to the gel before it is poured. This is a dye that stains DNA and when the gel is placed on a UV transilluminator, after electrophoresis, the fragments of DNA can be seen and compared with each other. This pattern is sometimes called a DNA fingerprint.

In this exercise you will compare the DNA of three bacterial viruses: phage lambda, phage ϕX 174, and virus X (which is either lambda or ϕX 174). The restriction enzyme Dra I is used to cut the DNA. It recognizes the base sequences

TTTAAA

AAATTT

Lambda contains 13 sites that can be cut by this enzyme, but ϕX 174 has only two. Therefore we should be able to identify virus X as either ϕX 174 or lambda depending on the number and size of the pieces separated by gel electrophoresis.

Note: Students should practice using micropipettors and loading a practice gel before doing the exercise.

Definitions

Buffer solution. A salt solution formulated to maintain a particular pH.

Electrophoresis. A procedure used to separate components by electrical charge.

Restriction enzymes. Enzymes isolated from bacteria that are used to cut DNA at specific sequences.

Objectives

1. To understand the use of restriction enzymes to cut DNA into segments.
2. To learn how to separate different lengths of DNA using gel electrophoresis.
3. To understand the use of DNA fingerprinting to identify DNA.

References

Alcamo, E. I. *DNA technology: The awesome skill,* 2nd ed. San Diego: Harcourt Academic Press, 2001.

Nester et al. Microbiology: A human perspective, 4th ed., 2004. Chapter 8, Section 8.13 and Chapter 9.

Sambrook, J., Fritsch, E.F., and Maniatis, T. *Molecular cloning,* 2nd ed. Cold Spring Harbor Laboratory Press, 1989.

Materials

Lambda DNA

ϕX 174 DNA

Unknown phage (either lambda or ϕX 174)

DNA size standard

Restriction enzyme Dra I

Microfuge tubes or Eppendorf tubes

Micropipettors

Gel box

Agarose

TBE buffer for electrophoresis (TRIS-EDTA-Borate)

TE buffer

Ethidium bromide

Stop mix (includes tracking dye)

Water bath

Microfuge

UV transilluminator

Goggles

Spatula

Disposable gloves

Procedure

Warning: Ethidium bromide is a potent mutagen.

1. Pour an agarose gel. The amount of agar will depend on the size of the gel box. The usual concentration is 0.7% agar and it is dissolved in the buffer TBE. Add ethidium bromide at 15 µl/200 ml agarose gel.

2. Add sterile distilled water (as determined by the instructor) to a sterile Eppendorf tube and then 1μl 10X TE buffer. Buffer is added before the enzyme so that the conditions are immediately optimal for the enzyme.

3. Add lambda DNA to the tube. This is usually 0.5–1.0 μg/μml. The instructor will indicate how much you should add.

4. Add the restriction enzyme Dra I.

5. Repeat for φX 174 DNA and the unknown virus DNA.

6. Each tube contains:
   Distilled water to bring total volume to 10 μl including enzyme.
   10X TE buffer
   DNA
   Enzyme

7. Mix and incubate for 30 minutes in a 37° water bath.

8. Add stop mix (usually 1 μl of 10X). This not only contains dyes, but will stop the reaction and help the sample sink to the bottom of the well.

9. Place gel in a gel box. Cover the agarose gel with running buffer and load the samples into the wells. Include a size standard in one of the lanes. This is DNA of known length so that you can estimate the size of the cut DNA pieces of the phage.

10. Attach the electrodes to the gel. Attach the positive electrode at the bottom of the gel. DNA is negatively charged so it will migrate to the positive electrode. (Remember run to red.) Run the gel until the colored bands of the dyes are separated. These dyes travel at about the same speed as the DNA and indicate how far the DNA has migrated.

11. Transfer the gel to a transilluminator with a spatula while wearing gloves.

12. View on a UV transilluminator.
    Warning: It is essential to wear goggles so that the UV does not damage your eyes.

13. Photograph the gel or make a drawing of the bands. Compare the number of bands that result from cutting of the phage DNA with Dra I.

NOTES:

EXERCISE

34

Laboratory Report:
Identifying DNA with Restriction Enzymes

Results

1. Record your results in table 34.1.

Table 34.1   Results of Phage DNA Electrophoresis

	Lambda	φX 174	Unknown (Phage X)
Number of fragments			
About what size is the largest piece?			
About what size is the smallest piece?			

2. What is the probable identity of phage X?

Questions

1. How were you able to estimate the size of the DNA fragments?

2. What is the purpose of adding ethidium bromide to the agarose gel?

3. What is the purpose of adding a tracking dye to the DNA before adding to the well?

4. Would you have the same number of fragments from each phage if you used a different restriction enzyme? Why or why not?

EXERCISE

35
Identification of Bacteria Using the Ribosomal Data Project

Getting Started

Identifying and classifying bacteria has always been more difficult than identifying and classifying plants and animals. Bacteria have very little differences in their structure, and while they are metabolically extremely diverse, it has not been clear which of these characteristics are the most important for identification or grouping. For example, is nitrogen fixation more significant than anaerobic growth, or endospore formation more important than photosynthesis?

The ability to sequence the DNA of microorganisms has offered a new solution. The closer organisms are related to each other, the more similar their **DNA nucleotide sequences.** Certain genes common to all bacteria can be sequenced and compared. What genes should be compared? The DNA coding for a part of the ribosome, namely the **16S** ribosome portion, is a good choice. Ribosomes are critically important for protein synthesis and any mutation is quite likely to be harmful. Some mutations, however, are neutral or perhaps even advantageous and these mutations will then be part of the permanent genome. These sequences change very slowly over time and are described as highly conserved. The 16S rRNA DNA segment is found in all organisms (slightly larger in eukaryotes) with the same function so the sequences can easily be compared.

This approach has had exciting results. Some specific base sequences are always found in some organisms and not others. These are called **signature sequences.** Also, new relationships between bacteria could be determined by comparing sequences of organisms base by base by means of computer programs. The more the sequences diverge, the more the organisms have evolved from one another.

Perhaps even more useful, these sequences can be used to identify bacteria. The sequences of at least 16,000 organisms are in public databases. If a new sequence is submitted to a database management computer, in seconds it will respond with the most likely identification of the species containing the sequence.

This exercise will give you a chance to send the DNA sequence of the 16S ribosomal RNA of an organism to the Ribosomal Database Project and determine the identification of the organism. Since many schools do not have the resources for determining the specific sequences, you will be given the sequences which you can enter on the Web and immediately receive an identification. You may also look for other sequences posted on the Web or in microbiological journals such as *Journal of Bacteriology* published by the American Society for Microbiology.

Definitions

DNA nucleotide sequence. The order that bases are found in a piece of DNA.

16S. S is an abbreviation for Svedberg. It is a unit of mass and measured by the rate a particle sediments in a centrifuge. The prokaryotic ribosome is made up of two main parts of 30S and 50S. The 30S particle is made up of the 16S rRNA + 21 polypeptide chains.

Signature sequences. DNA sequences of about 5–10 bases long found at a particular location in the 16S rRNA that are unique to Archaea, Bacteria, or Eukarya (eukaryotes).

Objectives

1. To understand the importance of the 16S rRNA sequence for the identification of organisms.
2. To understand how to identify organisms using the Ribosomal Database Project.

Reference

Nester et al. *Microbiology: A human perspective,* 4th ed., 2004. Chapter 9, Section 9.8.

Procedure

1. Open Netscape (or Explorer) on the computer.
2. Type in the URL http://rdp.cme.msu.edu
3. Click on "Online analysis."
4. Find "Sequence match" in purple column. Click on gray arrow in run column for the sequence match.
5. In the box "Cut and paste a sequence from your machine" at the bottom of the page, enter the sequence in the box. This probably would be most easily done by one person reading and another typing. Try typing in the first 150 bases as it may be sufficient.
6. Click on "Submit sequence" a few screens down. The parameters can be changed, but we will use the standard.
7. In just a few seconds the identification of the genus and species which most closely matches your sequence appears, including the percent of the sequences which matched your sample nucleotide sequence. Other species and their percent of similarity are also listed.

Hint: The organisms were studied in:

1. Exercise 32
2. Exercise 22
3. Exercise 23
4. Exercise 13 (genus only)
5. Photosynthetic bacteria (not found in any exercise)

Organism 1

```
   1    tctctgatgt tagcggcgga cgggtgagta acacgtggat aacctaccta taagactggg
  61    ataacttcgg gaaaccggag ctaataccgg ataatatttt gaaccgcatg gttcaaaagt
 121    gaaagacggt cttgctgtca cttatagatg gatccgcgct gcattagcta gttggtaagg
 181    taacggctta ccaaggcaac gatgcatagc cgacctgaga gggtgatcgg ccacactgga
 241    actgagacac ggtccagact cctacgggag gcagcagtag ggaatcttcc gcaatgggcg
 301    aaagcctgac ggagcaacgc cgcgtgagtg atgaaggtct cggatcgta aaactctgtt
 361    attagggaag aacatatgtg taagtaactg tgcacatctt gacggtacct aatcagaaag
 421    ccacggctaa ctacgtgcca gcagccgcgg taatacgtag gtggcaagcg ttatccggaa
 481    ttattgggcg taaagcgcgc gtaggcggtt ttttaagtct gatgtgaaag cccacggctc
 541    aaccgtggag ggtcattgga aactggaaaa cttgagtgca gaagaggaaa gtggaattcc
 601    atgtgtagcg gttaaatgcg cagagatatg gaggaacacc agtggcgaag gcgactttct
 661    ggtctgtaac tgacgctgat gtgcgaaagc gtgggaatca aacaggatta gataccctgg
 721    tagtccacgc cgtaaacgat gagtgctaag tgttaggggg tttccgcccc ttagtgctgc
 781    agctaacgca ttaagcactc cgcctgggga gtacgaccgc aaggttgaaa ctcaaaggaa
 841    ttgacgggga cccgcacaag cggtggagca tgtggtttaa ttcgaagcaa cgcgaagaac
 901    cttaccaaat cttgacatcc tttgacaact ctagagatag agccttcccc ttcgggggac
 961    aaagtgacag gtggtgcatg gttgtcgtca gctcgtgtcg tgagatgttg ggttaagtcc
1021    cgcaacgagc gcaaccctta gcttagttg ccatcattaa gttgggcact ctaagttgac
1081    tgccggtgac aaaccggagg aaggtgggga tgacgtcaaa tcatcatgcc ccttatgatt
1141    tgggctacac acgtgctaca atggacaata caaagggcag cgaaaccgcg aggtcaagca
1201    aatcccataa agttgttctc agttcggatt gtagtctgca actcgactac atgaagctgg
1261    aatcgctagt aatcgtagat cagcatgcta cggtgaatac gttcccgggt cttgtacaca
1321    ccgcccgtca caccacgaga gtttgtaaca
```

Organism 2

```
   1   taacacgtgg ataacctacc tataagactg ggataacttc gggaaaccgg agctaatacc
  61   ggataatata ttgaaccgca tggttcaata gtgaaagacg gttttgctgt cacttataga
 121   tggatccgcg ccgcattagc tagttggtaa ggtaacggct taccaaggca acgatgcgta
 181   gccgacctga gagggtgatc ggccacactg gaactgagac acggtccaga ctcctacggg
 241   aggcagcagt agggaatctt ccgcaatggg cgaaagcctg acggagcaac gccgcgtgag
 301   tgatgaaggt cttcggatcg taaaactctg ttattaggga agaacaaatg tgtaagtaac
 361   tatgcacgtc ttgacggtac ctaatcagaa agccacggct aactacgtgc
```

Organism 3

```
   1   gcctaataca tgcaagtaga acgctgagaa ctggtgcttg caccggttca aggagttgcg
  61   aacgggtgag taacgcgtag gtaacctacc tcatagcggg ggataactat tggaaacgat
 121   agctaatacc gcataagaga gactaacgca tgttagtaat ttaaaagggg caattgctcc
 181   actatgagat ggacctgcgt tgtattagct agttggtgag gtaaaggctc accaaggcga
 241   cgatacatag ccgacctgag agggtgatcg gccacactgg gactgagaca cggcccagac
 301   tcctacggga ggcagcagta gggaatcttc ggcaatgggg gcaaccctga ccgagcaacg
 361   ccgcgtgagt gaagaaggtt ttcggatcgt aaagctctgt tgttagaaga gaatgatggt
 421   gggagtggaa aatccaccaa gtgacggtaa ctaaccagaa agggacggct aactacgtgc
 481   cagcagccgc ggtaatacgt aggtcccgag cgttgtccgg atttattggg cgtaaagcga
 541   gcgcaggcgg ttttttaagt ctgaagttaa aggcattggc tcaaccaatg tacgctttgg
 601   aaactggaga acttgagtgc agaaggggag agtggaattc catgtgtagc ggtgaaatgc
 661   gtagatatat ggaggaacac cggtggcgaa agcggctctc tggtctgtaa ctgacgctga
 721   ggctcgaaag cgtggggagc aaagaggatt agataccctg gtagtccacg ccgtaaacga
 781   tgagtgctag gtgttaggcc ctttccgggg cttagtgccg gagctaacgc attaagcact
 841   ccgcctgggg agtacgaccg caaggttgaa actcaaagga attgacgggg gcccgcacaa
 901   gcggtggagc atgtggttta attcgaagca acgcgaagaa ccttaccagg tcttgacatc
 961   ccgatgcccg ctctagagat agagttttac ttcggtacat cggtgacagg tggtgcatgg
1021   ttgtcgtcag ctcgtgtcgt gagatgttgg gttaagtccc gcaacgagcg caaccccctat
1081   tgttagttgc catcattaag ttgggcactc tag
```

Organism 4

```
   1   ggtaccactc ggcccgaccg aacgcactcg cgcggatgac cggccgacct ccgcctacgc
  61   aatacgctgt ggcgtgtgtc cctggtgtgg gccgccatca cgaagcgctg ctggttcgac
 121   ggtgttttat gtaccccacc actcggatga gatgcgaacg acgtgaggtg gctcggtgca
 181   cccgacgcca ctgattgacg ccccctcgtc ccgttcggac ggaacccgac tgggttcagt
 241   ccgatgccct taagtacaac agggtacttc ggtggaatgc gaacgacaat ggggccgccc
 301   ggttacacgg gtggccgacg catgactccg ctgatcggtt cggcgttcgg ccgaactcga
 361   ttcgatgccc ttaagtaata acgggtgttc cgatgagatg cgaacgacaa tgaggctatc
 421   cggcttcgtc cgggtggctg atgcatctct tcgacgctct ccatggtgtc ggtctcactc
 481   tcagtgagtg tgattcgatg cccttaagta ataacgggcg ttacgaggaa ttgcgaacga
 541   caatgtggct acctggttct cccaggtggt taacgcgtgt tcctcgccgc cctggtgggc
 601   aaacgtcacg ctcgattcga gcgtgattcg atgcccttaa gtaataacgg ggcgttcggg
 661   gaaatgcgaa cgtcgtcttg gactgatcgg agtccgatgg gtttatgacc tgtcgaactc
 721   tacggtctgg tccgaaggaa tgaggattcc acacctgcgg tccgccgtaa agatggaatc
 781   tgatgttagc cttgatggtt tggtgacatc caactggcca cgacgatacg tcgtgtgcta
 841   agggacacat tacgtgtccc cgccaaacca agacttgata gtcttggtcg ctgggaacca
 901   tcccagcaaa ttccggttga tcctgccgga ggccattgc
```

Organism 5

1   agagtttgat cctggctcag agcgaacgct ggcggcaggc ttaacacatg caagtcgaac
61   gggcgtagca atacgtcagt ggcagacggg tgagtaacgc gtgggaacgt accttttggt
121   tcggaacaac acagggaaac ttgtgctaat accggataag cccttacggg gaaagattta
181   tcgccgaaag atcggcccgc gtctgattag ctagttggtg aggtaatggc tcaccaaggc
241   gacgatcagt agctggtctg agaggatgat cagccacatt gggactgaga cacggcccaa
301   actcctacgg gaggcagcag tggggaatat tggacaatgg gcgaaagcct gatccagcca
361   tgccgcgtga gtgatgaagg ccctagggtt gtaaagctct tttgtgcggg aagataatga
421   cggtaccgca agaataagcc ccggctaact tcgtgccagc agccgcggta atacgaaggg
481   ggctagcgtt gctcggaatc actgggcgta aagggtgcgt aggcgggttt ctaagtcaga
541   ggtgaaagcc tggagctcaa ctccagaact gcctttgata ctggaagtct tgagtatggc
601   agaggtgagt ggaactgcga gtgtagaggt gaaattcgta gatattcgca agaacaccag
661   tggcgaaggc ggctcactgg gccattactg acgctgaggc acgaaagcgt ggggagcaaa
721   caggattaga taccctggta gtccacgccg taaacgatga atgccagccg ttagtgggtt
781   tactcactag tggcgcagct aacgctttaa gcattccgcc tggggagtac ggtcgcaaga
841   ttaaaactca aaggaattga cgggggcccg cacaagcggt ggagcatgtg gtttaattcg
901   acgcaacgcg cagaacctta ccagcccttg acatgtccag gaccggtcgc agagacgtga
961   ccttctcttc ggagcctgga gcacaggtgc tgcatggctg tcgtcagctc gtgtcgtgag
1021 atgttgggtt aagtcccgca acgagcgcaa ccccgtcct tagttgctac catttagttg
1081 agcactctaa ggagactgcc ggtgataagc cgcgaggaag gtgggatga cgtcaagtcc
1141 tcatggccct tacgggctgg gctacacacg tgctacaatg gcggtgacaa tgggaagcta
1201 aggggtgacc cttcgcaaat ctcaaaaagc cgtctcagtt cggattgggc tctgcaactc
1261 gagcccatga agttggaatc gctagtaatc gtggatcagc atgccacggt gaatacgttc
1321 ccgggccttg tacacaccgc ccgtcacacc atgggagttg gctttacctg aagacggtgc
1381 gctaaccagc aatggggggca gccggccacg gtagggtcag cgactggggt gaagtcgtaa
1441 caaggtagcc gtaggggaac ctgcggctgg atcacctcct t

EXERCISE

Laboratory Report: Identification of Bacteria
Using the Ribosomal Data Project

Results

Organism	Bacteria or Archaea
1 _____	_____
2 _____	_____
3 _____	_____
4 _____	_____
5 _____	_____

Questions

1. What is an advantage of identifying an organism by using the Ribosomal Data Project?

2. What is a disadvantage?

3. It is generally thought by microbiologists that you cannot randomly create a sequence that the database will identify as an organism. Can you prove them wrong?

*I*NTRODUCTION to the Individual Projects

Many students taking a laboratory class in microbiology are planning careers in medicine and health-related fields. Therefore the emphasis in many courses is on laboratory exercises that will provide vital skills and concepts to prepare students for those occupations. The laboratory exercises tend to emphasize the control and identification of pathogenic organisms in addition to understanding basic microbiological principles.

Pathogenic organisms, however, make up a very small percentage of the known microorganisms that exist in the world. For the most part, it is the nonpathogenic prokaryotes that are responsible for recycling animal and plant material and are absolutely essential for making life on earth possible.

How are they able to carry out these activities? Prokaryotes have an astonishing versatility. The physiological abilities of prokaryotes make the eukaryotes appear very limited. For instance, no eukaryotes can fix nitrogen, oxidize sulfur for energy, produce methane and very few can grow anaerobically. The unique physiological tricks that bacteria use in breaking down and synthesizing molecules as well as their adaptation to their niche are very interesting. For instance there are organisms that specialize in taking one-carbon compounds produced by plants and building them into cellular material that can then be utilized by other organisms. Others break down complex molecules to simpler components to be incorporated by other bacteria. Some bacteria have adaptive strategies such as producing light or resisting the effects of radiation that are clearly important to the organisms' survival, but their roles are not understood.

Once you sample the world of microbiology, you may find it especially exciting and perhaps would like to investigate an organism on your own as an individual project. The following section presents some protocols for isolating bacteria that are particularly interesting for their unique physiological abilities. Some of these organisms are not easily hunted down and take patience and persistence to isolate. A large part of the project is finding the necessary equipment and source of samples. But if you enjoy a challenge these individual projects can be very rewarding. Your instructor will decide whether the results should be presented in a written paper, a poster, a report to the class, or both a written and an oral presentation.

Isolating these organisms is also a true investigation. The best source of some of these organisms has not been determined. It would be very helpful for future students attempting to isolate these organisms if a folder for each organism were kept in the laboratory. You could then add information to the folder about what you learned from your experience of isolating and identifying your particular organism.

The four projects include petroleum (hydrocarbon) degrading bacteria, luminescent bacteria, methylotrophs, and the UV resistant *Deinococcus*. They are arranged by the difficulty of their isolation, the hydrocarbon degraders being the most reliably successful.

NOTES:

INDIVIDUAL PROJECT 36

Hydrocarbon-Degrading Bacteria, Cleaning Up After Oil Spills

Getting Started

The biosphere contains a great variety of organisms that are collectively capable of breaking down just about all naturally occurring carbon compounds. This is fortunate because otherwise any compound not reduced to simpler molecules would accumulate in the environment. Sometimes it is useful to isolate the specific bacteria that are responsible for breaking down a particular compound. This can be done by using a medium which contains only that compound as the source of carbon. If you wanted to isolate an organism, for instance, that oxidizes phenol you could use a mineral salts medium for the nitrogen, phosphate, and sulfur and then use phenol for the carbon source.

Oil spills of crude petroleum are a serious threat to the marine environment. Several methods of removing the oil from the environment were tried in a serious oil spill in Alaska. One of the more useful was simply adding fertilizer to provide a source of nitrogen and phosphorus to encourage the growth of bacteria already present in the environment.

The hydrocarbons in oil are natural compounds found almost everywhere in nature. They are made up of a mix of different lengths of carbon chains saturated (covered) with hydrogen, called alkanes, and other carbon-hydrogen compounds. Since petroleum is formed by fossil plants, these same organic molecules are found in garden soil produced by growing plants. The same kinds of organisms that are present in marine environments breaking down seaweed (and potentially petroleum) are also present in the garden soil. Therefore even though muddy marine soil would be an ideal source for these organisms, garden soil is also an excellent source of hydrocarbon degraders.

Choose any petroleum product such as kerosene or fuel oil for the carbon source. Avoid products containing detergents or other additives.

Enrichment

The first step in these kinds of isolations is enrichment as a way of increasing the numbers of the desired bacteria. This means adding inoculum consisting of a source of the organism to a mineral medium with the chosen carbon source as the only source of carbon. In this case, soil would be added to the mineral medium, plus a hydrocarbon. The soil itself contains nutrients which can support the growth of many organisms, but the bacteria that are able to also use the hydrocarbon will have an advantage. They will be able to grow after the carbon compounds added with the soil have been exhausted.

1. For the enrichment prepare mineral salts medium broth (MSM). See table 36.1. Add 200 ml to a 500-ml flask and cover with a foil cap. Prepare several flasks to maximize your chances of successfully isolating the organisms.

Table 36.1 Minimal Salts Medium (Modified from E. Rosenberg)

NaCl	2.5 g (28.4g NaCl for marine organisms)
K$_2$HPO$_4$	4.74 g
KH$_2$PO$_4$	0.56 g
MgSO$_4$ · 7H$_2$0	0.50 g
CaCl$_2$ · H$_2$0	0.1 g
NH$_4$NO$_3$	2.5 g
Tap water	1 liter
Agar (for plates)	20 grams (15 grams Difco agar)
pH 7.1	

Note: 2.06 (NH$_4$)$_2$SO$_4$ and 3.15 KNO$_3$ can be substituted for 2.5 g NH$_4$NO$_3$

2. Add 0.1% vol/vol of the hydrocarbon of the carbon source (0.2 ml hydrocarbon/200 ml broth). Remember %vol/vol = ml/100 ml.
3. Add about a gram of soil to the broth.

4. Incubate the flask for two weeks at room temperature. These organisms are usually obligate aerobes, so shaking the flask is helpful, but not necessary. Avoid light to discourage the growth of algae. Since they fix CO_2, the algae can grow on CO_2 from the air without degrading the hydrocarbon.

Selection

The second step is selection. After enriching for the organism in a broth culture, your chances of isolating them are greatly increased. Streak the broth on mineral salts agar plates and add the hydrocarbon as described. The medium is very selective because only organisms able to utilize the hydrocarbon can grow.

1. Pour mineral salts agar plates. If you use plastic petri plates, check to see if the hydrocarbon you plan to use dissolves them. If it does, use glass plates.
2. Add the hydrocarbon carbon source such as fuel oil. (Since some petroleum products do not dissolve well in water, it was not placed in the agar.) Instead, invert the agar plate and place a piece of filter paper in the lid of the petri. Add 0.5 ml of the hydrocarbon to the filter paper and replace the inverted top of the agar plate (figure 36.1). The organisms will be able to grow on the fumes.
3. Streak the MSM agar plates with a loopful of inoculum from your enrichment flasks to obtain isolated colonies.

4. Prepare a control by adding water to the filter paper in another plate. This will control for organisms that might be able to degrade agar to obtain carbon and therefore would grow using agar instead of the hydrocarbon.
5. Incubate the plates at room temperature for about a week.

Isolation

1. Examine your plates carefully. Is there more than one type of colony? Note if any organisms are growing on the control plate—these must be using either nutrients contaminating the minerals in the medium or the agar itself (a few bacteria can utilize agar as a carbon source).
2. Purify your isolates by restreaking the organisms on a similar plate using the same hydrocarbon. Repeat until you can be sure you have at least one pure culture.
3. Make a Gram stain for an initial identification of the organism or organisms growing on the plates. If a Gram-negative rod is present, determine if it is oxidase positive. *Pseudomonas* is oxidase positive and well known for the ability to degrade unusual compounds. There are many Gram-positive organisms as well as other Gram-negative bacteria that are also involved in degrading complex molecules. It may not be possible for you to specifically identify your organism, but you can propose possible genera.
4. You might try different hydrocarbons to determine whether your organism can degrade other petroleum products. Does it grow on standard laboratory media? Can you think of other experiments using your isolated organism?

Reference

Rosenberg, E. *The hydrocarbon oxidizing bacteria in the Prokaryotes*, 2nd ed., Vol. I, pgs. 446–456. Albert Balows et al. eds. New York: Springer-Verlag, 1992.

Figure 36.1 The hydrocarbon is placed on the filter paper in the inverted petri dish.

Agar

Filter paper

INDIVIDUAL PROJECT 37

Luminescent Bacteria: Bacteria That Produce Light

Getting Started

Some marine bacteria have the ability to emit light, a process called *bioluminescence*. Most of these organisms live in relationship with tropical fishes. In one group of fish a specialized organ near their eye supports the growth of the bacteria. Covering and uncovering the organ with a special lid permits the fish to signal other fish in the darkness of deep water. Angler fish have a luminescent organ they dangle in front of them like a lure. Smaller fish are attracted to the light and are eaten. In the northern waters the fish do not have special organs for *luminescent* bacteria. Here these bacteria can be found in fish intestines, associated with squid, or free-living in salt water.

The enzyme bacterial luciferase requires both oxygen and a protein called an inducer. One organism alone cannot produce enough of the inducer to permit light production. A population of bacteria is needed to produce the critical amount of inducer and permit all the bacteria to produce light. This phenomenon is called *quorum sensing* and has recently been found in many other bacterial systems. Light production is very energy intensive—the equivalent of 6 to 60 molecules of ATP are needed for each photon of light produced. This system prevents one single bacterium from wasting energy to produce light that could not be detected.

It is clear that luminescence may be an advantage to bacteria living symbiotically with fish. In exchange for light they have a protected and rich environment. Squid also may have special organs which are associated with these bacteria. Some squid store bacteria in a sac and expel the bacteria when attacked, similar to a cloud of ink. Others may use the bacteria to produce light so that they do not cast a shadow on bright moonlit nights, thereby protecting them from predators.

More puzzling is the relationship of luminescent bacteria living in the gut of the fish. Since this is an anaerobic environment no light is produced, even though there is sufficient inducer. One proposal, which has little evidence to support it, is that the bacteria prefer the rich environment of the intestine, and when cast out into the water, try to return by producing light and inducing fish to swallow them. If they grow on a piece of detritus or organic material, they might accumulate enough inducer to emit light. In this case a population of bacteria would be more successful than a single cell.

Many marine dinoflagellates luminesce when the salt water is warm. They frequently can be seen at night when water is agitated by waves lapping on the beach or by an anchor thrown into the water. It is not known what role luminescence plays in their ecology either.

Procedure for Isolating Luminescent Bacteria

1. Prepare several plates of sea water complete agar (table 37.1). No one universal medium supports the growth of all luminescent bacteria, but this has been very successful.

Table 37.1	Sea Water Complete Agar
Sea water	750 ml
Glycerol	3.0 ml
Peptone	5.0 grams
Yeast extract	0.5 grams
Agar	15.0 grams
Water	250 ml
pH 7.5	

Note: The sea water can be actual sea water or any kind of artificial sea water made up in the lab or purchased from an aquarium supply store. (From K.H. Nealson)

2. Obtain a whole saltwater fish or purchase a squid at a seafood store. Fresh squid is best, but frozen is also good if it is not thawed with hot water. Results with frozen fish have not been very successful.

3. Dip a swab in the intestinal content of the fish and swab a third of the agar plate. Similarly cut the squid open and try swabbing various organs. It is not clear where the luminescent bacteria are in highest concentration, but try swabbing the outside surface and the interior. Dispose the swab in a wastebucket and continue streaking for isolation with a loop.

4. Incubate plates at room temperature 12–48 hours.

5. Observe for luminescent colonies. Take the plate in a dark room and permit your eyes to adjust for a few minutes. Circle any glowing colonies with a marking pen on the bottom of the plate. Try to look at the plate frequently in the first 24 hours because sometimes the bacteria are luminous for only a few hours. If your first attempt is unsuccessful, try again. Not all squid or fish have luminescent bacteria or at least ones that grow on this medium. If you leave the fish or squid at room temperature unwrapped, you will sometimes see glowing colonies on their surface the next day.

6. Restreak a glowing colony onto another plate to obtain a pure culture. The organism is probably either *Photobacterium* or *Vibrio*.

7. If you would like to demonstrate the luminescence of your organisms to the class, you can write a message or draw an image on an agar plate with a swab dipped in your pure culture.

References

Bergey's Manual of Systematic Bacteriology, Vol. 1. Edited by Krieg, Noel, and John G. Holt, Baltimore/London: Williams & Wilkens, 1984.

Nealson, K. H. "Isolation, identification and manipulation of luminous bacteria," pp. 153–166 in *Methods in enzymology*, Vol. 57. Edited by M. Delucca. New York: Academic Press, 1978.

INDIVIDUAL PROJECT 38

Methylotrophs, Organisms That Grow on One-Carbon Compounds

Getting Started

It may not seem very remarkable that bacteria can utilize methanol or other one-carbon compounds, but this source of nutrition actually presents special problems for bacterial metabolism. All of the compounds that the bacterium requires must be built from one-carbon precursors. Bacteria capable of this feat are called methylotrophs (meh thī´ low trofs). They should not be confused with methanogens or methane producers, which are members of the Archaea that synthesize the gas methane anaerobically.

Recently it was found that a reliable place to look for methylotrophs was on the underside of green leaves. Evidently plants continually produce a one-carbon compound, and the bacteria are able to utilize it and grow on the surface of leaves. If you press a leaf on an agar plate and then remove it, the pattern of the leaf can be seen in the resulting bacterial growth.

It is not necessary to enrich for these organisms growing on plants because the leaves already support a concentrated population. The medium is highly selective because only organisms able to utilize methanol will be able to grow.

Note: Another environment where methylotrophs may be found is in the traffic dust on the side of the road. These organisms may be important in degrading residual hydrocarbons found in vehicular exhaust.

Procedure

1. Prepare mineral salts + methanol + cyclohexamide agar plates (see "Materials"). Prepare mineral salts agar without a carbon source. Autoclave the agar, cool to 50°C and then add the methanol (the source of carbon) and the cyclohexamide before pouring plates. Methanol is a one-carbon alcohol and would evaporate in the autoclave. Cyclohexamide is an antifungal antibiotic. There are some yeasts (a fungus) that grow on leaves and also can utilize one-carbon compounds, so it is important to add cyclohexamide to prevent their growth. Antibiotics are usually added to the melted, cooled agar just before pouring plates because most antibiotics are heat sensitive.
Also prepare a control plate without the methanol.

2. Press the underside of a leaf on the agar and immediately remove. Do not use waxy leaves. Keep a record of the kind of leaves you use. Try several different kinds on separate plates.

3. Incubate at room temperature for 3–7 days and look for pigmented colonies. Are the colonies growing in the pattern of the leaf? Many methylotrophs are pink. Make a Gram stain of your isolate to make a tentative identification and to be sure you have bacteria and not yeast. Yeast are much larger than bacteria (about 10 μm in diameter) and appear purple in a Gram stain.

4. Restreak your organism on the same medium to obtain a pure culture.

5. After you have isolated your organism, try to determine what other carbon sources it can utilize in addition to methanol (if indeed it can).

6. There are two types of methylotrophs. One group can utilize ethyl alcohol or even sugars in addition to one-carbon compounds. The other group are restricted to methane and other one-carbon compounds. They are called methanotrophs.

If your organism is in the second group, report it as a methanotroph. If it is in the first group, it could be many different organisms including *Bacillus*, *Pseudomonas*, or *Vibrio*. Although you may try different tests, it is difficult to identify newly isolated organisms from the environment, so report the shape and Gram reaction.

Materials

Mineral Salts Medium

$(NH_4)_2SO_4$	1 gram
K_2HPO_4	7 grams
KH_2PO_4	3 grams
$MgSO_4.7H_2O$)	0.1 grams
Trace elements (if available)	1 ml
Tap water	1 liter
Agar (Difco)	15 grams
pH 7.0	

Note: Any kind of trace elements can be tried.

Autoclave at 120°C for 20 minutes.

Agar can be stored until ready to use. Before pouring plates melt agar heating in a container of boiling water 30 to 45 minutes. Cool to 50°C, add the methanol and cyclohexamide and then pour the plates. Pour the plates a day before you plan to use them so the surface will be dry.

After autoclaving and before pouring plates add the following:

20 μg/ml cyclohexamide or 1.0 ml of a 100 × stock solution/100 ml agar

0.1% methanol (0.1 ml methanol/100 ml) autoclaved, cooled agar before pouring plates

Stock Solutions

Antibiotics are required in very small amounts. The usual procedure is to make up a concentrated stock solution and then add it to the melted and cooled agar before pouring plates.

Cyclohexamide Stock Solution

A 100 × solution would be 100 × 20 μg/ml or 2,000 μg/ml. Since 2,000 μg is 2 milligrams, a stock solution would be 2 mg/ml. If you prepared 100 ml of a stock solution, you would add 200 mg (0.2 grams) to 100 ml water.

References

Madigan, M., Martinko, J., and Parker, J. *Brock biology of microorganisms*, 9th ed. Upper Saddle River, N.J.: Prentice Hall, 2000.

Perry, J. J., and Staley, J. T. *Microbiology dynamics and diversity*. Fort Worth: Harcourt Brace College Publishers, 1997.

INDIVIDUAL PROJECT 39

Deinococcus, Bacteria with Out-of-This-World Capabilities

Getting Started

Deinococcus are aerobic Gram-positive cocci that are truly remarkable for their ability to repair damage to their DNA. They are so resistant to radiation (including ultraviolet [(UV)] and gamma radiation) that they can survive radiation at least ten times more intense than other non-spore-forming organisms. In fact, they can survive radiation much more intense than can be found on earth. They are also very resistant to drying. These characteristics are very puzzling and *Deinococcus* is currently being studied to determine its DNA repair mechanisms. Mutations, caused by changes in DNA, are responsible for birth defects and many forms of cancer, so it would be very exciting to find ways of preventing mutations. Since this organism is so resistant to radiation, it is also being studied for a possible role in the biodegradation of toxic compounds in nuclear waste sites, a serious problem in many parts of the world.

A reliable source for isolation of these organisms has not been found. It has been isolated sporadically from a variety of environments such as creek water, soil, and air. One student held an open agar petri plate out the window while a friend drove down a freeway, and actually was successful in isolating it. Others have tried the same method and failed to isolate *Deinococcus*. A reasonable possible source is the hair from the back of a cow or a cow's tail because the back of a cow is constantly exposed to UV.

These organisms were previously named *Sarcina*, then *Micrococcus radiodurans* and now *Deinococcus*. The prefix *dein* means "strange."

Enrichment Procedure

In an enrichment procedure the organism that you are trying to isolate is encouraged to grow while other organisms are discouraged. In this enrichment almost all organisms will grow, but hopefully enough *Deinococcus* will be present to be seen on the irradiated plate.

1. Obtain some hair from the back or tail of a cow or any other source you think has possibilities.
2. Incubate the hair in tryptone yeast extract glucose (TYEG) broth (table 39.1) at least three days at 30°C. The organism seems to require growth factors in yeast extract. It grows slowly so incubate at least three days. The optimal temperature for growth is 30°C and therefore it grows slower above or below this temperature.

Table 39.1 TYEG Medium

Tryptone	5.0 grams
Yeast extract	3.0 grams
Glucose	1.0 gram
Tap water	1 liter
Agar	15 grams

Selection

1. Place 0.1 ml of the enrichment culture on a TYEG medium agar plate and spread with a swab or a bent glass rod. This will result in a lawn of bacterial growth. *Escherichia coli* is very sensitive to UV and *Bacillus* spores are very resistant, but not as resistant as *Deinococcus*. How can you prepare control plates of *E. coli* and *Bacillus* spores to compare with your *Deinococcus* enrichment plate? (Hint: See exercise 12.)
2. Irradiate the plates 1–3 hours under a UV lamp. See exercise 12 for precautions for working with a UV lamp. **WARNING:** Always wear safety goggles to protect your eyes and **never** look directly at the light.

3. Incubate several days at 30°C. Again, *Deinococcus* seems to have a decided preference for 30°C. The *Bacillus* plates can also be incubated at 30°, but *E. coli* prefers 37°, although it will grow at 30°.

4. After about 48 hours check the control plates—all the *E. coli* cells should be killed but the *Bacillus* spores are more resistant and may or may not be killed. Examine the other plate for possible colonies of *Deinococcus*. The colonies are frequently pigmented, usually yellow, orange, or red. A Gram stain of the colonies should show Gram-positive cocci arranged in packets of four and eight. How would a Gram stain help you distinguish *Deinococcus* from *Bacillus*?

5. Restreak your *Deinococcus* isolate on TYEG to obtain a pure culture. See color plate 28. Recommended identification is based on 16S rRNA analysis, so therefore you will have to rely on morphology of the organism as seen in the Gram stain and its resistance to radiation.

6. Once you have isolated *Deinococcus* perhaps you can think of other experiments you might try with it.

References

Balows, A. et al. *The prokaryotes*, 2nd ed., Vol. 4, p. 3736. New York: Springer-Verlag, 1992.

Bergey's Manual of Systematic Bacteriology, Vol. 2, p. 1035. Baltimore: Williams & Williams, 1986.

APPENDIX

Living Microorganisms (Bacteria, Fungi, Protozoa, and Helminths) Chosen for Study in This Manual

Acinetobacter, aerobic Gram-negative rods or coccobacilli in pairs. Commonly found in soil. Low virulence, can be an opportunistic pathogen. Naturally competent and therefore easily transformed.

Amoeba proteus, a unicellular protozoan that moves by extending pseudopodia.

Aspergillus niger, a filamentous black fungus with a foot cell, columella, and conidia.

Bacillus cereus, a Gram-positive rod, forms endospores, found in soil.

Bacillus subtilis, a Gram-positive rod, forms endospores, found in soil.

Candida albicans, an oval, budding, opportunistic dimorphic yeast.

Clostridium sporogenes, a Gram-positive rod, forms endospores, obligate anaerobe, found in soil.

Diphtheroids, Gram-positive irregular club-shaped rods.

Dugesia, a free-living flatworm.

Enterobacter aerogenes, a Gram-negative rod, coliform group, found in soil and water.

Enterococcus faecalis, a Gram-positive coccus, grows in chains, found in the intestinal tract of animals, occasionally an opportunist pathogen.

Escherichia coli, a Gram-negative rod, facultative, found in the intestinal tract of animals, coliform group, can cause diarrhea and serious kidney disease.

Escherichia coli, K-12 strain, commonly used in research. Host strain for λ phage.

Halobacterium salinarium, member of the Archaea, can live only in high salt solutions where they can utilize chemical energy from organic material but can also obtain energy from light.

Klebsiella pneumoniae, Gram-negative rod, coliform group, opportunist pathogen.

Micrococcus luteus, Gram-positive obligate aerobe, cocci are arranged in packets of four or eight. Part of the normal flora of the skin. The yellow colonies frequently seen as an air contaminant.

Moraxella, Gram-negative coccus found in normal flora.

Mycobacterium smegmatis, acid-fast with a Gram-positive type of cell wall.

Paramecium, a ciliated protozoan often found in pond water.

Penicillium species, a filamentous fungus with metulae, sterigmata, and conidia, also a source of antibiotics.

Propionibacterium acnes, Gram-positive irregular rod, obligate anaerobe, component of the normal skin flora living in sebaceous glands.

Proteus, Gram-negative rod, swarms on agar, hydrolyzes urea, can cause urinary tract infections.

Pseudomonas aeruginosa, Gram-negative rod, obligate aerobe, motile, opportunist pathogen, can degrade a wide variety of compounds.

Rhizopus nigricans, a filamentous fungus with stolons, coenocytic hyphae and a sporangium containing asexual sporangiospores.

Saccharomyces cerevisiae, eukaryotic fungal yeast cell, replicates by budding. Important in bread, beer and wine making, and the study of fungus genetics.

Spirillum volutans, Gram-negative curved organism with tufts of flagella at each pole, found in pond water and hay infusions.

Staphylococcus aureus, Gram-positive coccus, a component of the normal skin flora, but can cause wound infections, food poisoning, and toxic shock syndrome.

Staphylococcus epidermidis, Gram-positive coccus, a component of the normal skin flora.

Streptococcus mutans, normal flora of the mouth, forms gummy colonies when growing on sucrose.

Streptococcus pneumoniae, lancet-shaped Gram-positive cells arranged in pairs and short chains, pathogenic strains form capsules.

Streptococcus pyogenes, Gram-positive cocci in chains. The cause of strep throat, rheumatic fever, and glomerulonephritis.

APPENDIX

Dilution Practice Problems

See exercise 8 for an explanation for making and using dilutions.

1. If a broth contained 4.3×10^2 org/ml, about how many colonies would you expect to count if you plated:
   a. 1.0 ml
   b. 0.1 ml
2. Show three ways of making a:
   a. 1/100 or 10^{-2} dilution
   b. 1/10 or 10^{-1} dilution
   c. 1/5 or 2×10^{-1} dilution
3. Show two ways of obtaining a 10^{-3} dilution using 9.0 ml and 9.9 ml dilution blanks.
4. The diagram below shows a scheme for diluting yogurt before making plate counts. 0.1 ml was plated on duplicate plates from tubes B, C, and D. The numbers in the circles represent plate counts after incubation.
   a. Which plates were in the correct range for accurate counting?
   b. What is the average of the plates?
   c. What is the total dilution of tubes:

   A _____

   B _____

   C _____

   D _____

   d. How many organisms/ml were in the original sample of yogurt?
5. Suppose an overnight culture of *E. coli* has 2×10^9 cells/ml. How would you dilute it so that you would have countable plates? Diagram the scheme.

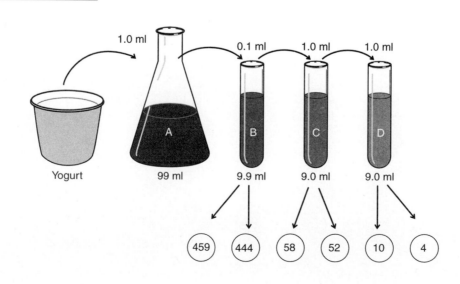

Answers:

1a. 430 colonies
1b. 43 colonies
2a. 1.0 ml into 99 ml
    0.1 ml into 9.9 ml
    10 ml into 990 ml
2b. 1.0 ml into 9.0 ml
    0.1 ml into 0.9 ml
    10 ml into 90 ml
2c. 1.0 ml into 4 ml
    0.1 ml into 0.4 ml
    10 ml into 40 ml

3.

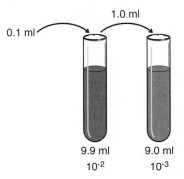

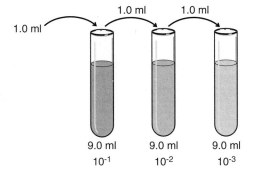

4a. Plates from tube C
4b. 55
4c. A: 10^{-2}
    B: 10^{-4}
    C: 10^{-5}
    D: 10^{-6}
4d. The number of colonies $\times$ 1/dilution $\times$ 1/sample on plate = number of organisms/ml

$$55 \times 1/10^{-5} \times 1/0.1 = 55 \times 10^{5} \times 10 = 55 \times 10^{6} \text{ or } 5.5 \times 10^{7} \text{ org/ml}$$

5.

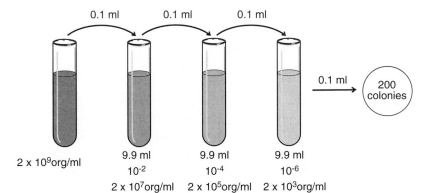

APPENDIX 3

Metric System, Use of with Conversions to the English System of Measurement

The metric system enjoys widespread usage throughout the world in both the sciences and nonsciences. Your studies in microbiology provide an excellent opportunity to learn how to use the metric system, particularly in the laboratory. The derivation of the word "metric" comes from the word "meter," a measure of length.

Some examples of where the metric system is used in your microbiology course are: media preparation, culture storage, and the measurement of cell number and size. The four basic measurements taken with the metric system are those concerned with weight, length, volume, and temperature. The beauty of using the metric system is that all of these measurements are made in units based on multiples of 10. Not so with the English system where measurements are made in units based on different multiples, e.g., 16 ounces in a pound, 12 inches in a foot, 2 pints in a quart, and 212°F, the temperature at which water boils. With the metric system, weight is measured in grams (g), length in meters (m), volume in milliliters (ml), and temperature in degrees Celsius (°C).

The prefixes of metric measurements indicate the multiple. The most common prefixes for metric measures used in microbiology are:

kilo (k) $= 10^3 = 1,000$

centi (c) $= 10^{-2} = 0.01$

milli (m) $= 10^{-3} = 0.001$

micro (μ) $= 10^{-6} = 0.000001$

nano (n) $= 10^{-9} = 0.000000001$

Unit Weight	Abbreviation	Equivalent
kilogram	kg	1000 g, 10^3 g
gram	g	1000 mg (0.035 ounces; 454 g = 1 pound)
milligram	mg	10^{-3} g
microgram	μg	10^{-6} g
nanogram	ng	10^{-9} g
picogram	pg	10^{-12} g
Length		
kilometer	km	1000 m (0.62 miles)
meter	m	100 cm (3.3 feet)
centimeter	cm	10^{-2} m (2.5 cm = 1 inch)
millimeter	mm	10^{-3} m
micrometer	μm	10^{-6} m (synonym = micron)
nanometer	nm	10^{-9} m
angstrom	Å	10^{-10} m
Volume		
liter	l	1000 ml (1.1 quarts; 3.8 liters = 1 gallon)
milliliter	ml	10^{-3} liter
microliter	μl	10^{-6} liter
Temperature		
Celsius	°C	0°C = 32°F, 100°C = 212°F
		To convert from °C to °F: (°C × 9/5) + 32
		To convert from °F to °C: (°F × 5/9) − 32

Note: See Appendix 2 for some related information.

APPENDIX

pH Adjustment of Liquid and Agar Growth Media

pH in essence is a measurement of the acidity of a solution in terms of its hydrogen ion (H^+) concentration. By examining table 1 you can see that it is measured on a scale of 0 to 14. It is commonly measured either colorimetrically with indicator dyes or electrometrically with a pH meter. The indicator dyes are either weak acids or weak bases which change their colors at a specific pH (see table 2). Usually when dyes are used to measure pH, a color comparison is made between the solution being measured and tinted pH buffer standards. Use of the standards helps increase accuracy. The accuracy of colorimetric pH measurements is not as accurate as electrometric measurements. At times it may be within 0.1 of a pH unit although errors up to 1.0 pH unit are possible. The test is usually conducted by transferring a small amount of growth medium into a clear white depression dish. A drop of indicator dye is added and the color compared with the tinted pH standards. It is generally sufficient for preparing growth media allowing a short range in the initial medium pH. It is more often used for preliminary pH adjustment of the culture medium. Other factors reducing accuracy of the colorimetric method are variations in temperature of the solution being measured, and the presence of salts and other colloidal particles such as pigments, all of which may cause color variations.

When using the pH meter for measuring the pH of growth media, it becomes important to do the following in order to obtain reliable measurements:

1. Solutions to be measured must be well mixed and free of temperature fluctuations.

2. pH electrodes should be rinsed well with distilled water before using—thereby removing any contaminants.

Some principles related to pH determination are shown in table 1. As you can see, the pH of pure water is 7, the pH on the scale which is neutral. Below 7 is acid and above 7 is basic. Also note that the greater the H^+ ion concentration, the lower the pH of that solution. The lower the pH the greater the acidity of that solution. The numbers are exponential; thus a solution with a pH of 6 has 10 times as many H^+ ions as a solution with a pH of 7. At a pH of 7 and above the reverse occurs, thus a solution with a pH of 8 has 10 times as many hydroxyl (OH^-) ions as a solution with a pH of 7. The amount of acid or hydroxyl ions required to change the pH depends on the nature of the solution (e.g., is it weakly buffered or strongly buffered?). What point on the pH scale is best for growth of bacteria and fungi? Most growth media for bacteria range between pH 6.5 and 7, whereas most media for fungi range between pH 5.5 and 6. For both groups, the growth response is not detrimental for a small change in growth medium pH; however, if incubated over a period of time or generations, a pronounced change in growth medium pH is apt to affect culture viability.

After weighing the necessary amount of growth medium, add most of the solvent (distilled water), mix well, and measure the pH of the dissolved solution. If the pH is alkaline (above 7) and the required pH is 6.5, it becomes necessary to lower the pH. A 1 molar solution of hydrochloric acid (HCl) is usually used. The H^+ ions combine with the excess hydroxyl (OH^-) ions in the growth medium to form H_2O, thus neutralizing them, thereby making the growth medium more acidic. Periodically measure the pH of the medium, adding small amounts of 1 molar HCl until the desired pH is attained. Next add the required amount of water necessary to make the final volume of growth medium.

Table 1	Growth of Microbes as It Relates to Range on the pH Scale

Optima: Most bacteria, pH 7; Most fungi, pH 5.5

1 2 3 4 5 **6 7** 8 9 10 11 12 13 14

Acidic   Neutral   Basic

← Increased acidity   Increased alkalinity →

Conversely, if the initial pH of the medium is too acid, it becomes necessary to raise the pH of the medium. A 1 molar sodium hydroxide solution is usually used. The hydroxyl (OH^-) ions combine with the excess hydrogen (H^+) ions in the growth medium forming water, thereby neutralizing them, causing the growth medium pH to become more alkaline. Next add the required amount of water necessary to make the final volume of medium and autoclave at S.T.P.T.

Note: When preparing a solid growth medium, after adjustment of the pH, place the flask of medium on a hot plate, add a stirring bar, initiate stirring and while heating **slowly,** add the agar, a small amount at a time. Once in solution cap the flask and autoclave at S.T.P.T

Most microbiology laboratories use sulfonophthalein-based indicators for preliminary colorimetric adjustment of culture medium pH, or as indicators of microbial metabolism in the media themselves (see *Difco Manual*, 9th ed., 1953, p. 295).

These indicators are nontoxic in the amounts used in media. Table 2 shows some of the more commonly used indicators.

| Table 2 | Sulfonophthalein Colorimetric Indicators Showing Preparation, pK, pH Range, and Metabolic Color Changes |

| | | **Full Colors** | | | |
Indicator	P*	pH Range	pK	Acid	Alkali
Bromcresol green	14.3	3.8–5.4	4.67	Yellow	Blue
Bromcresol purple	18.5	5.2–6.8	6.3	Yellow	Purple
Brom thymol blue	16.0	6.0–7.6	7.0	Yellow	Blue
Phenol red	28.2	6.8–8.4	7.9	Yellow	Red
Cresol red (alkaline)	26.2	7.2–8.8	8.3	Yellow	Red
Meta cresol purple (alkaline)	26.2	7.4–9.0	8.32	Yellow	Purple
Thymol blue (alkaline)	21.5	8.0–9.6	8.9	Yellow	Blue

*P: The ml of 0.01 N NaOH required per 0.1 g of indicator. Dilute to 250 ml with distilled water to prepare a 0.04% solution for use as an indicator for colorimetric pH determinations.

APPENDIX

Use of the Ocular Micrometer for Measurement of Relative and Absolute Cell Size

Determination of cell dimensions is often used in microbiology where it has numerous applications. Examples include measurement of changes in cell size during the growth cycle, determining the effect of various growth factors on cell size, and as a taxonomic assist in culture identification. Measurements are made by inserting a glass disc with inscribed graduations (see figure 2a), called an ocular micrometer, into the ocular of the microscope (See figure 1).

It is not necessary to calibrate the ocular micrometer to determine the *relative* size of cells. For this purpose one can examine either a wet mount or stained preparation of cells with the ocular micrometer. By measuring the length in terms of number of ocular micrometer divisions, you might conclude that cell X is twice as long as cell Y. For such purposes determination of cell length in absolute terms, such as number of micrometers, is not necessary.

For determining the *absolute* size of cells it becomes necessary to first measure the length in micrometers (μm) between two lines of the ocular micrometer. For this purpose a stage micrometer with a scale measured in micrometers becomes necessary for μm calibration of the ocular micrometer. The stage micrometer scale (see figure 2b) is such that the distance between two lines is 0.01 mm (equivalent to 10 μm). By superimposing the stage micrometer scale over the ocular micrometer scale, one can determine absolute values in microns between two lines on the ocular micrometer scale. The absolute value obtained is also dependent on the objective used. For example, with the low power objective, seven divisions on the ocular micrometer = one division on the stage micrometer. Thus with the low power objective, one division on the ocular micrometer = .01 mm/7 = 1.40 micrometers (μm).

Procedure for Insertion, Calibration, and Use of the Ocular Micrometer

1. Place a clean stage micrometer in the mechanical slide holder of the microscope stage.

2. Using the low power objective, center and focus the stage micrometer.
3. Unscrew the top lens of the ocular to be used, and then carefully place the ocular micrometer with the engraved side down on the diaphragm inside the eyepiece tube (see figure 1), followed by replacing the top lens of the ocular.
   Note: With some microscopes the ocular micrometer is inserted in a retaining ring located at the base of the ocular.
4. To calibrate the ocular micrometer, rotate the ocular until the lines of the ocular micrometer are superimposed over the lines of the stage micrometer.
5. Next move the stage micrometer until the lines of the ocular and stage micrometer coincide at one end.
6. Now find a line on the ocular micrometer that coincides precisely with a line on the stage micrometer.
7. Determine the number of ocular micrometer divisions and stage micrometer divisions where the two lines coincide. A case in point is shown in figure 2c.

A relatively large practice microorganism for determining average cell size, with both the low and high power objectives, is a yeast cell wet mount. You must first calibrate the ocular micrometer scale for both objectives using the stage micrometer. Next determine the average cell size of a group of cells, in ocular micrometer units, using both the low and high power objectives. Attempt to measure the same group of cells with both objectives. You may find it easier to first measure them with the high power objective, followed by switching to the low power objective. You will have mastered the technique if you obtain the same average cell size answer with both objectives.

Note: Care should be taken when inserting and removing the ocular micrometer from the ocular.

Figure 1 Location of the ocular and stage micrometers.

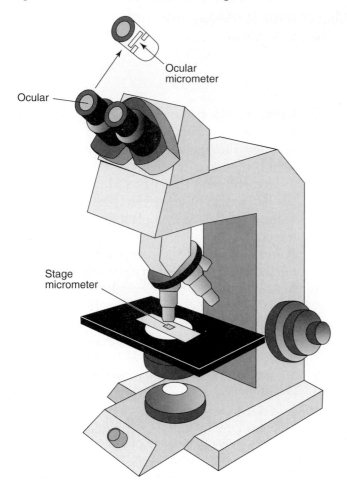

Ocular
micrometer

Ocular

Stage
micrometer

Before inserting make certain the ocular micrometer is free of dust particles by cleaning both sides
with lens paper moistened with a drop of lens
cleaning solution. Install and remove only in an
area free of air currents. After removing the ocular
micrometer reexamine the ocular for dust particles.
If present consult your instructor.

Figure 2 Calibration of the ocular micrometer.

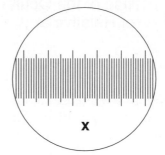

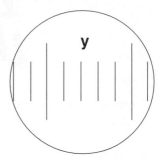

(a) Ocular micrometer
The diameter (width)
of the graduations
in microns must be
determined for
each objective.

(b) Stage micrometer
The graduations
are .01mm
(10 µm) wide.

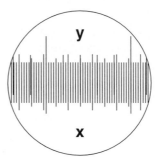

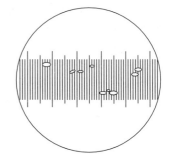

**(c) Superimposing of ocular
micrometer scale (x) over
the stage micrometer scale (y)**
Note that seven divisions of the
ocular micrometer equal one
division of the stage micrometer
scale (.01 mm).

One division of $x = \dfrac{.01}{7}$

$= .0014$ mm $= 1.4$ µm

(d) Based on the Figure 2(c)
calculations, what is the
average length in microns
of the rod shaped bacteria?

APPENDIX

Use of the Hemocytometer for Determining Total Cell Number in a Liquid Suspension

A method faster than the plate count (see exercise 8) for determining the total number of cells present in a liquid suspension is one in which the hemocytometer is used in conjunction with the microscope. With this method an aliquot of suspended cells is introduced between a cover glass suspended on mounts above the hemocytometer counting chamber (figure 1). The liquid depth between the cover glass and the counting chamber is 0.1 millimeter (mm).

The counting chamber is divided into a series of small squares in which the smallest squares are 1/400 of a square mm (see the central large square of figure 2). Thus a square mm would contain 400 small squares. The central large square is surrounded by double lines in order to make it easier to visualize when counting cells.

The hemocytometer is difficult to use with small cells because the thickness of the hemocytometer is such that it can only be used with the low and high power objectives, thereby making it difficult to distinguish individual small cells. For cells such as white blood cells, yeasts, and larger bacterial cells it is sometimes quite useful. It is used routinely for doing white blood cell counts and often for following the course of cell growth and multiplication in a liquid medium. For learning purposes yeast is an excellent test organism.

1. Dilute a test tube suspension of yeast such that clouding is barely visible with the naked eye. You may need to further dilute the sample if you find the individual cells too dense to count with the hemocytometer.
2. Wash the hemocytometer and hemocytometer cover glass with soapy water, rinse with distilled water, and dry the cover glass and hemocytometer counting surface with lens paper or Kimwipes. Make certain that all oily residues are removed from these areas.
3. Place the cover glass over the counting chamber area.
4. Tighten the test tube cap of the yeast suspension and shake thoroughly.
5. Using a Pasteur pipet or plastic dropper remove approximately 0.3 ml, and controlling the flow with your forefinger, place the tip in the V-shaped indention of the counting chamber adjacent to the edge of the cover glass (see figure 1a).

Figure 1 (a) Top view of hemocytometer showing sample introduction point. (b) Side view of hemocytometer showing the cover glass rests on the cover glass mounting supports and the distance (0.1 mm) between the top counting surface of the hemocytometer and the underside of the cover glass.

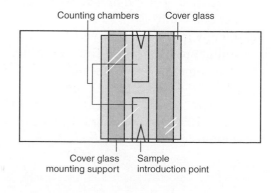

(a) Top view

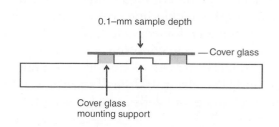

(b) Side view

Figure 2 Ruling of a hemocytometer showing the subdivisions of a central square millimeter. The central square is surrounded by double lines.

$\frac{1}{400}$ sq mm $\frac{1}{25}$ sq mm

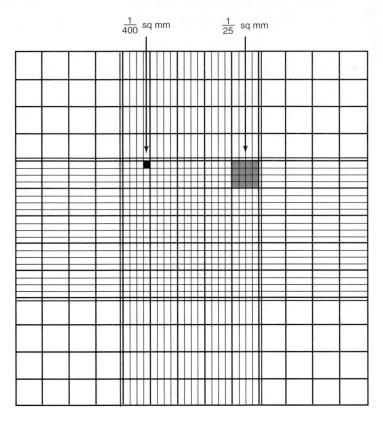

6. Slowly let the counting chamber fill by capillarity, making sure that the suspension does not go between the cover glass and cover glass mounting supports of the counting chamber (see figure 1b). Such an error will raise the height of the fluid under the cover glass which needs to be exactly 0.1 mm. If such an event occurs, return to step 2.

7. With care, place the hemocytometer on the stage of the microscope such that the hemocytometer counting chamber is centered underneath the low power objective.

8. Focus and make a total count of the number of cells in a predetermined number of small squares of the double-lined central area of the hemocytometer, for example, 100 small squares. For ease in counting, a budding yeast cell should be treated as one cell. Assuming you find 500 cells in 100 small squares, what is the total number of cells per ml of sample?

9. Calculations:
   a. 100 small squares = 1/4 of a square mm; in which instance, 500 cells × 4 = 2,000 cells per square mm.
   b. As previously mentioned the liquid depth of the counting chamber is 0.1 mm. Thus in order to determine the number of cells per cubic mm of fluid, it becomes necessary next to multiply by a factor of 10 in order to obtain the number of cells per cubic mm. Thus 2,000 cells/sq mm × 10 = 20,000 cells per cubic mm.
   c. Finally in order to convert a cubic millimeter (mm) to a cubic milliliter (ml), it becomes necessary to multiply by a factor of 1,000 because 1,000 cu mm = 1 cu ml. Thereby 20,000 cells per cu mm × 1,000 = 20×10^6 yeast cells per ml of the original suspension.

APPENDIX

Preparation of Covered Slide Cultures for Study of Intact Structure of a Mold Thallus

In exercise 19 you no doubt discovered that rapid growing molds such as *Penicillium*, *Aspergillus*, and *Rhizopus*, when cultivated on an agar plate, tend to spread thickly over the plate, making it difficult to distinguish morphological characteristics of the intact vegetative thallus. One way of sometimes improving this situation is to use a weaker growth medium containing a lower level of carbohydrates and a lower pH, for example, Sabouraud's medium in which the glucose concentration is reduced from 20 g to 10 g/l and the pH is reduced from pH 5.6 to 4.0. Doing so will slow down the growth rate. It may also distort normal growth. Another way, perhaps, to reach beyond this goal is to use a covered slide culture. You may wish to try both methods. An added suggestion you may wish to try is to incorporate the weaker growth medium with use of the covered slide culture.

One method for preparing a covered slide culture is as follows:

1. Place two pieces of filter paper on the bottom of a glass petri dish. Place a U-shaped piece of bent glass tubing, approximately 2–3 mm in diameter, on the filter paper, followed by placing a glass microscope slide on the piece of glass tubing (the tubing raises the slide above the filter paper). Add a microscope cover slip, approximately 15 mm square, and sterilize the petri dish and its contents.

2. Prepare and sterilize a flask of Sabouraud's dextrose agar or whatever agar you decide to use, and after cooling to approximately 40°C (so called "cheek temperature") aseptically pour some of the agar into a second sterile plastic petri dish.
   Note: Make the depth of the agar no more than 1–2 mm deep. Two mm = 0.08 inches. If too deep you will not be able to use the high dry objective.

3. Sterilize a spatula by flaming, and aseptically cut the agar into cubes about 15 mm square.

To help simplify this operation inscribe a pattern of a petri dish bottom on a sheet of paper. Next, with a dark marking pencil, draw parallel straight lines 15 mm apart from one another. Turn the pattern 90 degrees and draw a second set of parallel straight lines 15 mm apart from one another. The net result is a template with sixteen or more 15 mm cubes. Set the petri dish bottom on the template, aseptically remove the cover, and with a sterile spatula slice the agar along the template lines. You now have sufficient cubes of agar for use in preparing additional moist chambers.

4. Dip the end of a spatula in alcohol, flame to sterilize, and then aseptically remove a cube of agar and place it near the center of the glass slide contained in the sterile petri dish.

5. With a sterile loop aseptically remove some mold from an agar slant culture, and lightly inoculate the agar cube on the edges of all four sides.

6. With sterile forceps gently place the coverslip on top of the agar square.

7. Carefully add enough sterile water to moisten the filter paper, seal the edges of the petri dish with sealing tape to prevent drying, and incubate the petri dish at 25°–30°C for 3 to 6 days.

8. Examine visually for growth along the edges of the agar cube. If not seen reincubate the petri dish. If seen remove the slide and observe with the low and high power objectives of the microscope. Take care when focusing the high power objective in that the thickness of the agar may be such that the lens is apt to come in contact with the coverslip. Often, because of the moisture in the covered chamber, excellent disentangled fruiting structures can be seen along the edges of the coverslip.

Note: Should you have difficulty with seeing intact fruiting bodies with the preceding method, you may wish to try the Henrici method. This method is particularly valuable for untangling sporangiophore entanglements, especially with genera such as *Rhizopus* and *Mucor*.

1. A large clean cover glass (24 × 40 mm) is flamed and placed in a sterile petri dish. Next a drop of sealing wax is deposited on each end. With a hot spatula spread the wax out to form a layer approximately 5 mm wide and less than 2 mm thick across the cover glass ends.

2. A clean slide is now heated in the Bunsen flame and placed in a sterile petri dish containing a sheet of filter paper. Next the sterile cover glass is aseptically transferred to the central area of the slide with the cement side down. With a hot spatula the cement is softened so that it will adhere, not so hot that it will liquify and run. One should now have a culture chamber arranged as shown in figure 1, with a space a little less than 2 mm deep between the cover glass and the slide.

Figure 1

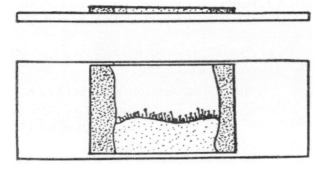

Courtesy of Helen Mitchell.

Figure 2

Courtesy of Helen Mitchell.

3. Melt a tube of Sabouraud's agar, cool to a temperature your cheek will tolerate, and inoculate the agar with spores of the mold you wish to study.

4. With a sterile capillary pipet transfer some of the agar to one side of the slide such that it covers an area similar to that shown in figure 1.

5. Moisten the filter paper with sterile water, carefully seal the bottom of the petri dish with sealing tape, and incubate at room temperature, approximately 25 to 30 degrees Celsius.

6. Examine after 3 to 6 days for evidence of growth. When seen remove the slide from the moist chamber and examine for growth (see figure 2).

Reference

Skinner, C., Emmons, C., and Tsuchiya, H. *Henrici's molds, yeasts, and actinomycetes*, 2nd ed. New York, London: John Wiley & Sons, Inc., 1947.

Media to Help You Master Microbiology!

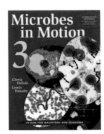

MICROBES IN MOTION 3 CD-ROM
By Gloria Delisle and Lewis Tomalty
ISBN 0-07-233438-X

This interactive, easy-to-use, general microbiology CD-ROM helps students actively explore and understand microbial structure and function through audio, video, animations, illustrations, and text. Eighteen modules cover topics from microbial genetics to vaccines.

HYPERCLINIC 2 CD-ROM
by Lewis Tomalty and Gloria Delisle
ISBN 0-07-232312-4

Evaluate realistic case studies that include a patient history and description of signs and symptoms. Students can either analyze the results of physician-ordered clinical tests to reach a diagnosis or evaluate a case study scenario and then decide which clinical samples should be taken and which diagnostic tests should be run. More than 200 pathogens are profiled, 105 case studies presented, and 46 diagnostic tests are covered.

MICROBIOLOGY: A HUMAN PERSPECTIVE ONLINE LEARNING CENTER
www.mhhe.com/nester4

Explore this dynamic site designed to help you get ahead and stay ahead in your study of microbiology. Some of the activities you will find on the website include:

- Self-quizzes to help you master material in each chapter.

- Animations to assist you in visualizing difficult microbiology concepts.

- Flash cards to ease your learning of new vocabulary.

- Clinical case studies to allow you to apply your knowledge in realistic situations.

McGraw-Hill Higher Education

ISBN 0-07-247624-9

90000

9 780072 476248

www.mhhe.com